Uchechukwu Opara

Hypersurface Theory in the Resolution of Differential Equations

Uchechukwu Opara

Hypersurface Theory in the Resolution of Differential Equations

Second Edition

LAP LAMBERT Academic Publishing

Publisher:
LAP LAMBERT Academic Publishing
is a trademark of
Dodo Books Indian Ocean Ltd. and OmniScriptum S.R.L publishing group

120 High Road, East Finchley, London, N2 9ED, United Kingdom
Str. Armeneasca 28/1, office 1, Chisinau MD-2012, Republic of Moldova, Europe
Managing Directors: Ieva Konstantinova, Victoria Ursu
info@omniscriptum.com

Printed at: see last page
ISBN: 978-3-659-66039-9

Zugl. / Approved by: Abuja, African University of Science and Technology, Diss.,2014

The ever-present importance of differential equations in mathematics and scientific modeling at large cannot be overemphasized. Despite the growing repository of techniques for tackling differential equations (D.E.'s), the concept of hypersurfaces is rather ubiquitous in their formulations - particulary in the case of Partial Differential Equations (P.D.E.'s). For instance, we may consider the nth order Dirichlet boundary value problem;

$$\begin{cases} \sum_{|\alpha| \leq n} D^\alpha f & = & g \quad \text{in } \Omega \\ f & = & h \quad \text{on } \partial\Omega \end{cases} \qquad (P_1)$$

or the nth order Neumann boundary value problem;

$$\begin{cases} \sum_{|\alpha| \leq n} D^\alpha f & = & g \quad \text{in } \Omega \\ \dfrac{\partial f}{\partial N} & = & h \quad \text{on } \partial\Omega \end{cases} \qquad (P_2)$$

for multi-indices α; where $\Omega \subset \mathbb{R}^m$ is open, bounded and uniformly of class C^1. The set $\partial\Omega$ in either setting is a hypersurface in $\mathbb{R}^m$, and of course the boundary conditions are significant in determining well-posedness and the nature of any solution f of (P_1) or (P_2). In (P_2), N is recognized as the Gauss map of $\partial\Omega$.

Some of the most potent tools in general D.E. theory such as trace theorem and Stokes' theorem often involve the concept of hypersurfaces. In Ordinary Differential Equations (O.D.E.'s), scoping the depth of the method

i

of solving by symmetry considerations can give rise to hypersurfaces (as Lie groups) which ought to be well-understood when using this approach. Symmetry properties, when present, are applicable to P.D.E.'s (P_1) and (P_2) as well, for appreciable simplification of the problems.

The points stated above serve to motivate the choice of the study of hypersurfaces in this regard, which incorporates tools from standard Riemannian geometry. It should be mentioned that all possible relevant content on the general subject here is inexhaustible, so the included content is designed to elucidate crucial extracts for further research. The reader of this concise text will unmistakenly detect the limitations of delving into D.E.'s without any hint of the presented geometric theories.

It also should be mentioned that background information acquired from tertiary pure mathematical studies will ease the assimilation and usefulness of the text content to the reader. Particulary, some knowledge of O.D.E.'s, P.D.E.'s, multivariable calculus, differential geometry, abstract algebra, functional analysis and convex analysis is needed. The author does not consider any section of the text to be fully drawn out at length, but recommends it as an interface between tertiary studies in mathematics and a basis for advanced research in related areas.

Contents

iv

Applications of Hypersurface Theory in the Resolution of Classical Differential Equations

Uchechukwu Michael Opara

African University of Science and Technology, Abuja, Nigeria
Mathematics Department.

November, 2014

v

Introduction

This body of work builds up diverse concepts related to the overall theme.

In the first chapter, intrinsic properties of hypersurfaces are discussed, with the fore-knowledge that many arising problems of interest are cast in the intrinsic setting for the reference manifolds involved. For contrast, extrinsic properties of hypersurfaces are discussed in the same chapter to draw up a more thorough description. In the second chapter, Riemannian geodesics are explained at considerable length. In the third chapter, we discuss a method for simplifying differential equations in the form of symmetry groups. In the concluding fourth chapter, there are practical illustrations of the applications of concepts presented in the first and third chapters. They are cast as analyses of variational problems.

Dirichlet's problem and the Ricci flow are motivated as extensions for the content in the fourth chapter, drawing on theoretical value from all the previous chapters.

CHAPTER 1

INTRINSIC VS. EXTRINSIC PROPERTIES OF HYPERSURFACES

Preview. Intrinsic properties of submanifolds of linear spaces must be understood in order to study them independent of the ambient spaces. We will focus on Riemannian hypersurfaces of Euclidean spaces here; pointing out the most essential contrasts between intrinsic and extrinsic properties and also certain striking links between them.

1.1 Introduction.

A $m-$manifold-without-boundary M^m is a set for which each point in M^m has a neighborhood in M^m that is diffeomorphic to an open subset of $\mathbb{R}^m$. Intrinsic properties of a manifold are those which can be obtained only from the first fundamental form, while hypersurfaces are submanifolds of co-dimension 1 (one dimension less) in the ambient space. Let S be an analytic, oriented hypersurface embedded in a real n-dimensional Euclidean space (E), locally or globally parametrized by the diffeomorphism

$$
\begin{aligned}
f: \quad & U = \overset{\circ}{U} \subseteq \mathbf{R}^{n-1} \quad \longrightarrow \quad && S \\
& (x_1, x_2, \cdots, x_{n-1}) := u \;\longmapsto\; && f(u) = (y_1, y_2, \cdots, y_n) = p \, .
\end{aligned}
$$

The Riemannian metric tensor g of S is defined by the matrix of the first fundamental form $[g_{ij}]$, where $g_{ij} = \langle \frac{\partial f}{\partial x_i}, \frac{\partial f}{\partial x_j} \rangle$ for $1 \leq i, j \leq n - 1$.

More formally, the tensor g is represented by the tensor product notation

$$g = \sum_{i,j=1}^{n-1} g_{ij} dx_i \otimes dx_j.$$

For this reason, we are also right to say that all intrinsic properties of S are obtainable directly from the metric tensor. Alternatively, we may define intrinsic properties of any manifold as those which can be recognized tangentially at each point, without any reference to the orthogonal complement in the ambient space.

At this juncture, we can immediately infer that for any open set, all its meaningful properties are intrinsic; since the tangent space to an open set at any of its points coincides with the tangent space to the ambient space at that point. Reducing the dimension of targeted manifolds by one, we have hypersurfaces in the ambient space, which are far more didactic with regard to this topic than open sets which are of equal dimension with the ambient space. Uniqueness of the outward unit normal at each point for orientable, differentiable hypersurfaces also creates rich prospects for their exploration. We will proceed with relevant points about hypersurface curvatures, directional and covariant vector derivates; and differential forms.

1.2   Curvatures of Hypersurfaces

All notions of curvature are based on principal curvatures of manifolds. They are obtained from the matrix of the first fundamental form $[g_{ij}]$ as well as that of the second fundamental form $[h_{ij}]$. Consistent with previous notation, $h_{ij} = \langle \frac{\partial^2 f}{\partial x_i \partial x_j}, N \rangle$ and N is the Gauss map or outward unit normal of the hypersurface S.

The principal curvatures of S are the eigenvalues of its shape operator, defined also to be the negative differential of the Gauss map

$$-DN = [g_{ij}]^{-1}[h_{ij}] := [g^{ij}][h_{ij}].$$

For this reason, the $n-1$ dimensional manifold S has $n-1$ principal curvatures. By virtue of their reliance on the one-dimensional orthogonal complement spanned by N_p to the tangent space T_pS at each $p \in S$, the principal curvatures are extrinsic properties of S. This means we may alter the principal curvatures of a given manifold by changing the way it is embedded in the ambient space. The mean curvature of S equals the mean of all principal curvatures and thus it is given by the formula

$$H = \frac{Tr(-DN)}{n-1} .$$

With reference to the most important category of hypersurfaces which are orientable smooth surfaces in $\mathbb{R}^3$, we can identify that mean curvatures are also extrinsic because of surfaces with the same first fundamental form but different mean curvatures. For example, the plane and cylinder have the same first fundamental form, but for the plane $H = 0$ while for the cylinder $H \neq 0$.

The Gaussian curvature of S equals the product of all principal curvatures and thus it is given by the formula

$$K = det(-DN) = \frac{det[h_{ij}]}{det[g_{ij}]} .$$

This quantity presents one striking link between extrinsic and intrisic properties of a given manifold in Gauss's *Theorema Egregium* or remarkable theorem. This theorem states that the Gaussian curvature K of a two-dimensional surface element $f : U = \overset{\circ}{U} \subseteq \mathbf{R}^2 \rightarrow \mathbb{R}^3$ of class C^3 depends only on the first fundamental form of the surface, and thus is an intrinsic property of the surface. The remarkable aspect of this theorem is that $det[h_{ij}]$ can be obtained only from the matrix $[g_{ij}]$ - a result which can be extrapolated generally to higher dimensional hypersurfaces as well[1] (p.164). Further illustrations given below will permit us to relate and contrast between critical intrinsic and extrinsic properties of hypersurfaces, ensuing with the forum of directional and covariant vector derivatives.

1.3   Vector Field Derivation

Let Y be a differentiable vector field defined on some open set of $\mathbb{R}^n$ and X a fixed directional vector at some point p of the open set; which is to say $X(p) \in T_p\mathbb{R}^n$. Then the directional derivative of Y in the direction of X

exists and is given by $D_X Y|_p := DY(X)|_p = \lim\limits_{t \to 0} \dfrac{1}{t}\left(Y(p+tX) - Y(p)\right).$
DY here represents the Jacobi matrix. The vector-valued partial derivatives
of Y; $D_{e_i} Y = \dfrac{\partial Y}{\partial y_i}$ correspond to the cases of $X = e_i$ with $e_1, e_2, \cdots, e_n$ being
the canonical basis vectors for $\mathbb{R}^n$. Consequently, for a general vector field
$X = \sum\limits_{1 \le i \le n} X^i e_i$, we have the result

$$D_X Y|_p = \sum\limits_{1 \le i \le n} X^i D_{e_i} Y .$$

Similarly, for our differentiable manifold S parametrized by $f : U = \overset{\circ}{U} \subseteq$
$\mathbb{R}^{n-1} \longrightarrow S \subset \mathbb{R}^n$, given a differentiable vector field Y along S and X a
tangent vector field to S at each point $p \in f(U)$; the directional derivative
vector field $D_X Y|_p$ is well-defined along S for each $p \in f(U)$. It is essential
also to take derivatives in the directions of the parameters $x_1, \cdots, x_{n-1}$ of
U. The derivative of Y in the direction of x_i is the case of $X = \dfrac{\partial f}{\partial x_i}$ and in
particular $D_{\frac{\partial f}{\partial x_i}} \dfrac{\partial f}{\partial x_j} = \dfrac{\partial^2 f}{\partial x_i \partial x_j}$. This relation is especially important when
X and Y are both tangent vector fields to S, so that the set $\left(\dfrac{\partial f}{\partial x_i}\right)_{1 \le i \le n-1}$
comprises all basis vectors for either vector field. Hence, we often apply this
relation in computing covariant derivatives, which are defined only for tangent vector fields to a manifold.

If X and Y are differentiable tangent vector fields to S, then the expression

$$\nabla_X Y = D_X Y - \langle D_X Y, N \rangle N$$

is called the covariant derivative of Y in the direction of X. Note that the
covariant derivative is the tangential portion of the ordinary extrinsic directional derivative, so that $\nabla_X Y$ is again another tangent vector field to S.
Being a tangential property, the covariant derivative is intrinsic to the hypersurface S. Consulting the Christoffel symbols of the second kind Γ_{ij}^k, we
realize that

$$\nabla_{\frac{\partial f}{\partial x_i}} \dfrac{\partial f}{\partial x_j} = \sum\limits_{1 \le k \le n-1} \Gamma_{ij}^k \dfrac{\partial f}{\partial x_k} .$$

The matrix of the first fundamental form $[g_{ij}]$ completely determines the
Christoffel symbols and as a result it completely determines all covariant

5

derviatives as well. In particular, the Christoffel symbols are obtained as

$$\Gamma_{ij}^{k} = \sum_{1 \leq l \leq n-1} \frac{1}{2} g^{kl} \left(\frac{\partial}{\partial x_i} g_{jl} + \frac{\partial}{\partial x_j} g_{il} - \frac{\partial}{\partial x_l} g_{ij} \right).$$

Indeed, the covariant derivative is often used to define other important intrinsic properties of S. For instance, for $n \geq 3$, the concept of geodesics of S is sufficiently explained using the covariant derivative. Let $c \subset S$ be a twice-differentiable one-dimensional submanifold or curve parametrized by arclength, then c is a geodesic of S if and only if $\nabla_{\dot{c}} \dot{c} \equiv 0$. Note that since the vector field $\dot{c}$ is tangent to a submanifold of S, then it is also tangent to S. In cases where straight lines are admitted by manifolds, such as the case of the entire Euclidean space, these are always geodesics because the Christoffel symbols Γ_{ij}^{k} vanish identically. Other instances of geodesics which can be immediately identified occur in 2-surfaces of revolution $(\theta, \phi) \mapsto (r(\theta) \cos \phi, r(\theta) \sin \phi, h(\theta))$, in which meridian curves given by $\phi = constant$ are always geodesics. On the other hand, curves given by $\theta = constant$ are geodesics only for values θ_0 for which $\dot{r}(\theta_0) = 0$.

Another intrinsic property defined using the covariant derivative is called the Riemann - Christofell curvature tensor, which we now briefly characterize. Let X, Y and Z be tangent vector fields to the hypersurface S. The curvature tensor field denoted $Rm(X,Y)Z$ is given by

$$Rm(X,Y)Z := \nabla_X \nabla_Y Z - \nabla_Y \nabla_X Z - \nabla_{(\nabla_X Y - \nabla_Y X)} Z.$$

Hence, the Riemann - Christoffel curvature tensor is a (1,3) - tensor type (that is to say, 3-linear with one vector argument in its co-domain) and it describes the tidal force experienced by a rigid body undergoing motion along geodesics of a manifold. On the other hand, covariant and directional vector derivation are (1,2) - tensor types.

Considered to be the most basic intrinsic quantifier of hypersurface curvature, the Ricci tensor ($C_1 R$ or Ric) is a $(0,2)$ tensor type obtained via the Riemann-Christoffel curvature tensor. It is defined as the first contraction of this curvature tensor, and given by the formula

$$(C_1 R)(Y, Z) = Tr(X \mapsto Rm(X, Y)Z) = \sum_{1 \leq i \leq n-1} \langle Rm(E_i, Y)Z, E_i \rangle$$

where $\{E_i\}_{1 \leq i \leq n-1}$ is an orthonormal basis for $T_p S$ at each $p \in S$. The Ricci tensor is symmetric, which is to say $Ric(Y, Z) = Ric(Z, Y)$. Its trace is

6

known as the *Ricci scalar* or scalar curvature which we denote R;

$$R := \sum_{1 \leq i,j \leq n-1} \langle Rm(E_i, E_j)E_j, E_i \rangle .$$

In the fundamental case of regular two-dimensional surfaces, it can be verified that the scalar curvature is given by twice the Gaussian curvature; $R = 2K$. More generally, the scalar curvature is always a function of the 2-sectional Gaussian curvatures of manifolds. The Gauss curvature $K(\sigma)$ of the 2-dimensional section σ of T_pS is defined by the following formula for any two linearly independent vectors $v_1, v_2 \in \sigma$,

$$K(\sigma) := K(v_1, v_2) = \frac{\langle Rm(v_1, v_2)v_2, v_1 \rangle}{\|v_1 \wedge v_2\|^2} .$$

The scalar term in the denominator,

$$\|v_1 \wedge v_2\|^2 = \|v_1\|^2 \|v_2\|^2 - \langle v_1, v_2 \rangle^2$$

may also be obtained by determining the corresponding covectors v_1^* and v_2^* (in E^*) to v_1 and v_2 by Riesz representation. Subsequently,

$$\|v_1 \wedge v_2\|_E = \frac{1}{\sqrt{2}} \|v_1^* \wedge v_2^*\|_{\Im^2(E)}$$

with the usual norm $\|\pi_i \wedge \pi_j\|_{\Im^2(E)} = \sqrt{2}$ $(i \neq j)$ for any orthonormal basis $\{\pi_i\}_{1 \leq i \leq n}$ of E^*. Furthermore, we have

$$\langle Rm(u,v)w, z \rangle = \frac{1}{6} \frac{\partial^2}{\partial s \partial t}(K(u + sz, v + tw) - K(u + sw, v + tz))|_{(s,t)=(0,0)}$$

so that the 2-sectional Gaussian curvatures determine the Riemann-Christoffel curvature tensor completely.

In the next comparison to be made between intrinsic and extrinsic hypersurface properties, we will focus on a special class of tensors called differential forms, which are crucial in the study of multivariate integration.

1.4 Intrinsic and Extrinsic Differential Forms

For an m-dimensional linear space V, the set of multilinear functions mapping from $V^k \to \mathbb{R}$ denoted by $\Im^k(V)$ constitutes a linear space of dimension m^k.

This linear space of (0, k) - tensor types is regularly just referred to as the set of k-tensors on V. The subspace of differential k-forms, denoted $\bigwedge^k(V)$, is made of the set of alternating multilinear k-forms. Thus, for any $\omega \in \bigwedge^k(V)$, we have that

$$\omega(v_1, \cdots, v_i, \cdots, v_j, \cdots, v_k) = -\omega(v_1, \cdots, v_j, \cdots, v_i, \cdots, v_k) \, ,$$

where the only indices switched are i and j for vectors $v_1, v_2, \cdots v_k \in V$. The subspace of differential k-forms on V is of dimension C_k^m and is empty for $k > m$. Also, the subspace of differential m-forms on V is of dimension 1. Let $\varphi_1, \varphi_2, \cdots, \varphi_m$ be a set of basis vectors for the dual space V^* of V, then any differential k-form $\omega \in \bigwedge^k(V)$ can be expressed in the wedge product or exterior product notation as $\alpha \varphi_{i_1} \wedge \cdots \wedge \varphi_{i_k}$ for a constant α and distinct integers $i_1, i_2, \cdots i_k$ between 1 and m. Observe that $V^* = \bigwedge^1(V)$.

Consider now the (n - 1)-dimensional hypersurface S of $\mathbb{R}^n$ parametrized initially, for which the tangent space to S at $p := T_pS$, is an (n - 1)-dimensional subspace of $\mathbb{R}^n_p$ for each $p \in S$. We are interested in the fibre bundle $\Omega^k(S)$, which equals $\bigcup_{p \in S} \bigwedge^k(T_pS)$. An exterior k - form on S is defined as a section of this fibre bundle and it is given by a finite sum of terms of the type

$$\omega^k = h \ \varphi_{i_1} \wedge \cdots \wedge \varphi_{i_k}$$

for a smooth function $h : S \to \mathbb{R}$. Here, $\varphi_1, \varphi_2, \cdots, \varphi_{n-1}$ is a formula for the set of basis vectors for the tangent dual bundle on S and the distinct integers $i_1, i_2, \cdots i_k$ lie between 1 and n - 1 . This is to say that ω_p^k is a differential k-form on the tangent space T_pS so that

$$\omega_p^k(v_1, v_2, \cdots v_k) = h(p) \ (\varphi_{i_1} \wedge \cdots \wedge \varphi_{i_k}(v_1, v_2, \cdots v_k))$$

for each $p \in S$ and vectors $v_1, \cdots v_k \in T_pS$.

Of course, the ambient space $\mathbb{R}^n$ is itself a differentiable manifold so we may attribute to it similar properties as given S above. First, the canonical basis vectors for the dual of the tangent space $\mathbb{R}^n_u$ for each $u \in \mathbb{R}^n$ are the collection of projection maps often denoted by $\{dy_i\}_{i=1}^n$. Hence, any differential k-form on $\mathbb{R}^n$ can be decomposed to a finite sum of terms

$$h \ dy_{i_1} \wedge \cdots \wedge dy_{i_k}$$

for smooth functions $h : \mathbb{R}^n \to \mathbb{R}$ and distinct integers $i_1, i_2, \cdots i_k$ between 1 and n.

8

The differential k-forms on $\mathbb{R}^n$ are the *extrinsic differential k-forms* on S while sections of the fibre bundle $\Omega^k(S)$ constitute the *intrinsic differential k-forms* on S. In the vein of multivariate integration along the manifold S, we are mostly interested in the differential (n - 1)-forms, as these are the only forms which may be integrated over the hypersurface to yield non-zero output. We may either integrate the differential (n - 1)-forms on $\mathbb{R}^n$ over S, or we may integrate the exterior (n - 1)-forms from the fibre bundle $\Omega^{n-1}(S)$. The (n - 1)-forms on $\mathbb{R}^n$ being extrinsic to S suggests that the result of integration over the hypersurface varies depending on how S is embedded in $\mathbb{R}^n$. In the following basic example, we see an illustration of this remark, after which we will study links between intrinsic and extrinsic differential forms along with certain useful consequences.

The winding 1-form on $\mathbb{R}^2$

Setting the two canonical projection maps on $\mathbb{R}^2$ as dx, dy dual to the canonical basis vectors e_1, e_2 for $\mathbb{R}^2$, the differential 1 - form called the winding form defined on the open subset $\mathbb{R}^2 - \{(0,0)\}$ is given by

$$\omega^1 = \frac{-y}{x^2 + y^2}dx + \frac{x}{x^2 + y^2}dy \ .$$

Integrating ω^1 over a closed curve surrounding the origin, we obtain a measure of how often the curve winds about the origin. For instance, when we integrate ω^1 over a single parametrization of the simple closed curve S^1 - the unit circle in $\mathbb{R}^2$ centered at the origin $\{(0,0)\}$, we get a result of 2π. When we translate S^1 in $\mathbb{R}^2$ by a constant vector v, we still get a result of integration as 2π for $\|v\| < 1$; but we get 0 for $\|v\| > 1$. This is because in the latter case, S^1 no longer encompasses the origin, although it did in the former. For this reason the differential one - form ω^1 is extrinsic to all embedded curves in $\mathbb{R}^2$, and we will make further reference to this form in due course.

1.4.1 Integration of Intrinsic Differential Forms

Recall that for each $p \in S$, the space $\bigwedge^{n-1}(T_pS)$ is of dimension 1. We also refer to the volume element of S as its hypersurface element and denote it by dS_p; and this is the canonical basis tensor field for the space $\bigwedge^{n-1}(T_pS)$. More specifically, we say that dS is a basis for any section of the fibre bundle $\Omega^{n-1}(S)$, meaning that any exterior (n - 1)-form on S can be written as hdS for a smooth function or zero - form $h : S \to \mathbb{R}$. Let $\{\alpha_1, \alpha_2, ..., \alpha_{n-1}\}$ be set as an orthonormal basis of functions for the tangent bundle of S, with $[\alpha_1(p), \alpha_2(p), ..., \alpha_{n-1}(p)]$ as the orientation for T_pS $\forall p \in S$. Then we obtain the corresponding basis elements for $(T_pS)^*$: $\varphi_1, \varphi_2, ..., \varphi_{n-1}$ such that $\varphi_i(\alpha_i) = 1$ and $\varphi_i(\alpha_j) = 0$ for $i \neq j$. In this event, the exterior product $\varphi_1 \wedge \varphi_2 \wedge ... \wedge \varphi_{n-1}$ gives the volume element dS on S.

Practically, the basis $\{\varphi_i\}_{1 \leq i \leq n-1}$ can be tedious and therefore not ideal to compute. As a useful alternative, dS is pulled back to U by the parametrization f. We state this transformation as follows.

$$f^* : \quad \bigwedge^{n-1}(T_pS) \quad \longrightarrow \quad \bigwedge^{n-1}(\mathbf{R}_u^{n-1})$$
$$dS \quad \longmapsto \quad f^*(dS) = f^*(\varphi_1 \wedge \varphi_2 \wedge ... \wedge \varphi_{n-1})$$

Properties we should recognize of f^*; the pullback of f for a differentiable function f, $f : \mathbb{R}^m \to \mathbb{R}^n$; $(x_1, x_2, \cdots, x_m) \mapsto (y_1, y_2, \cdots, y_n)$ are listed below.

1. $f^*(dy_i) = \sum_{j=1}^m D_j f^i dx_j = df^i$

2. $f^*(\omega_1 + \omega_2) = f^*(\omega_1) + f^*(\omega_2)$

3. $f^*(h \cdot \omega) = (h \circ f) f^* \omega$; for a functional $h : \mathbb{R}^n \to \mathbb{R}$

4. $f^*(\omega \wedge \eta) = f^* \omega \wedge f^* \eta$

5. $f^*(d\omega) = d(f^* \omega)$

The space $\bigwedge^{n-1}(\mathbf{R}_u^{n-1})$ is of dimension 1 so that any element from there must be a 0-form multiplied by the volume form on $(\mathbf{R}_u^{n-1})$ which is $dx_1 \wedge dx_2 \wedge \cdots \wedge dx_{n-1} \in \bigwedge^{n-1}(\mathbf{R}_u^{n-1})$. The 0-form required here is the square root of the determinant of the positive - definite matrix of the first fundamental form, so that $f^*(dS) = \sqrt{det[g_{ij}]} dx_1 \wedge dx_2 \wedge \cdots \wedge dx_{n-1}$. This means integrating the volume element dS over $f(U)$ is the same as integrating the differential (n - 1)-form $\sqrt{det[g_{ij}]} dx_1 \wedge dx_2 \wedge \cdots \wedge dx_{n-1}$ over U. This relation is relevant in integrating any intrinsic (n - 1)-form over the entire manifold due to the single dimension of $\bigwedge^{n-1}(T_pS)$ $\forall p \in S$. For a section hdS of the fibre bundle $\Omega^{n-1}(S)$, integrating this exterior (n - 1)-form over $f(U)$ is the same as integrating $(h \circ f) \sqrt{det[g_{ij}]} dx_1 \wedge dx_2 \wedge \cdots \wedge dx_{n-1}$ over U.

As such, we observe how the first fundamental form is used to determine the integration of intrinsic differential forms over hypersurfaces. Apart from the basic application of integrating volume forms, the first fundamental form is expediently engaged when integrating intrinsic differential forms over any manifold. Having pointed this out, we will now make mention of certain applicable relationships between intrinsic and extrinsic differential forms.

1.4.2 Links between Intrinsic and Extrinsic Differential Forms

Although the k-forms in $\Omega^k(S)$ do not belong in $\Omega^k(\mathbb{R}^n)$, the following statement gives an important way in which intrinsic differential forms on S are related to the ambient linear space.

Proposition - Each smooth exterior k-form $\omega^k \in \Omega^k(S)$ can be continuously extended to a smooth section of $\Omega^k(V)$ for some open set $V \supset S$ of $\mathbb{R}^n$.

Proof - We will refer to a standard theorem from finite-dimensional functional analysis for this proof. We set

$$\omega^k = h \; \varphi_{i_1} \wedge \cdots \wedge \varphi_{i_k}$$

for a smooth function $h : S \to \mathbb{R}$ and $\varphi_1, \varphi_2, \cdots, \varphi_{n-1}$ as a set of basis vectors for the tangent dual bundle on S and the k distinct integers $i_1, i_2, \cdots i_k$ between 1 and n - 1. By way of the analytic form of the Hahn - Banach theorem [2] (p. 39), each linear map on T_pS can be continuously extended to a linear map on $\mathbb{R}^n_p$ for each $p \in S$. Consequently, for the basis elements $\{\varphi_i\}_{i=1}^{n-1}$ of $(T_pS)^* = \bigwedge^1(T_pS)$, we have extensions into $(\mathbb{R}^n_p)^*$ which we denote respectively $\{\overline{\varphi_i}\}_{i=1}^{n-1}$. This means the exterior product

$$\overline{\varphi_{i_1}} \wedge \cdots \wedge \overline{\varphi_{i_k}}$$

is a differential k-form on the ambient space $\mathbb{R}^n$. By observing the pullback operator f^*, the formula for this form can be evaluated smoothly on $f(U)$ because we have the formula

$$f^*(\overline{\varphi_{i_1}} \wedge \cdots \wedge \overline{\varphi_{i_k}})(v_1, \cdots v_k) = \overline{\varphi_{i_1}} \wedge \cdots \wedge \overline{\varphi_{i_k}}(Df(v_1), \cdots Df(v_k))$$

for vectors $v_1, \cdots v_k \in \mathbb{R}^{n-1}_{f^{-1}(p)}$ and the (1 , 1)–tensor field Df is continuously differentiable on U.

Concerning the 0-form $h : S \to \mathbb{R}$, intrinsic and extrinsic differentiablity are equivalent so that its derivative must be taken in the classical sense. That is to say $\forall p \in S$, there exists an open subset of $\mathbb{R}^n$ containing p, say V_p, on which h is defined and differentiable. Taking

$$V = \bigcup_{p \in S} V_p \ ,$$

we get that $h \ \overline{\varphi_{i_1}} \wedge \cdots \wedge \overline{\varphi_{i_k}}$ is a differential k-form on the open set and submanifold V of $\mathbb{R}^n$. As such, the restriction of this k-form to S is ω^k and it is a section of the fibre bundle $\Omega^k(V)$, as required.

The above observation has a number of useful consequences, and we now proceed to highlight such an implication in a theorem of Stokes concerning manifolds with boundary. Stokes' theorem states;

' If M is a compact oriented k-dimensional manifold - with - boundary and ω is a (k - 1) form on M, then $\displaystyle\int_M d\omega = \int_{\partial M} \omega$ where ∂M is given the induced orientation.' [4] (p. 79)

We will use this theorem to show how the volume of an n-ball in $\mathbb{R}^n$ is related to the volume of its boundary. Let our manifold M be an n-ball centered at the origin with radius r, and ω be the hypersurface element on the boundary ∂M of M which is a (n - 1)-sphere. We cannot implement this theorem successfully if ω is to be taken intrinsically to ∂M because the intrinsic exterior differential of ω equals 0. We therefore use the extrinsic extension of ω to $\mathbb{R}^n$:

$$\overline{dS} := \omega = \sum_{1 \leq i \leq n} N^i (-1)^{i+1} dy_1 \wedge dy_2 \wedge \cdots \wedge \underline{dy_i} \wedge \cdots \wedge dy_n,$$

where $N : \partial M \to \mathbb{R}_p^n$; $p \mapsto (N^1(p), N^2(p), \cdots, N^n(p))$ is the Gauss map of the hypersurface ∂M for each $p \in \partial M$. Underlining the term dy_i indicates that it is excluded from the exterior product. The Gauss map in this case is simply obtained as

$$N : \partial M \to \mathbb{R}_p^n \ ; \ p \mapsto \left(\frac{y_1}{r}, \frac{y_2}{r}, \cdots, \frac{y_n}{r} \right) \ .$$

We then get

$$d\omega = \sum_{1 \leq i \leq n} \frac{(-1)^{i+1}}{r} dy_i \wedge dy_1 \wedge dy_2 \wedge \cdots \wedge \underline{dy_i} \wedge \cdots \wedge dy_n = \frac{n}{r} dV \ .$$

We use dV to connote the volume element on $\mathbb{R}^n$ which coincides with that of all open n-dimensional submanifolds, such as the interior of the n-ball currently under investigation. Note that the volume of our compact manifold equals the volume of its interior since the interior is non-empty. Finally, we get the well - known relation

$$\frac{n}{r} \int_{B^n} dV = \int_{S^{n-1}} dS \; ;$$

denoting the n-ball of radius r and its boundary the (n - 1)-sphere, respectively by B^n and S^{n-1}. In conclusion, we now discuss the important differential form properties of exactness and closedness.

Closed and exact differential forms

A form ω is closed if $d\omega = 0$ and exact if $\omega = d\eta$ for some form η. Every exact form is closed since if $\omega = d\eta$ then $d\omega = d(d\eta) = 0$. The converse does not necessarily hold. Poincaré's Lemma on this issue gives a sufficient condition for closed forms to be exact;

'Let $W \subset \mathbb{R}^n$ be an open set star-shaped with respect to the origin, then every closed form on W is exact.' [4] (p. 20)

A set is said to be star-shaped with respect to the origin if it includes the origin as well as the entire line segment connecting the origin to each of its other points. This theorem can only be meaningfully applied to extrinsic differential forms or extrinsic extensions of exterior forms on manifolds, because the intrinsic differentials of intrinsic forms of manifolds have no meaning on the outside ambient space. This is to be taken into consideration especially since our target manifolds which are hypersurfaces have empty interiors and often do not even possess any convex sections.

Referring back to the basic example of the winding 1-form (ω^1) on $\mathbb{R}^2 \backslash \{0\}$ given earlier, we have a differential form which is closed but not exact. If ω^1 were an exact 1-form, then it would be the differential of a smooth functional 0-form mapping from $\mathbb{R}^2 \rightarrow \mathbb{R}$ and also, its integral over any simple closed curve in $\mathbb{R}^2$ would vanish. In mathematical physics, exact one-forms bear the analogy to conservative force fields on simple closed curves since the integral of such vector fields over simple closed curves must always vanish. More specifically, for a conservative force field F and closed curve c, we have

$$\int_c F.ds = 0 \; .$$

13

For a one-form ω on $\mathbb{R}^2$ to be the differential of a smooth 0-form, say h, then given a simple closed curve parametrized precisely once by

$$\gamma : [a, b] \to \mathbb{R}^2 \; ; \; t \mapsto \gamma(t)$$

we have that,

$$
\begin{aligned}
\int_{\gamma} \omega &= \int_{[a,b]} \gamma^* \omega = \int_a^b \gamma^*(dh) \\
&= \int_a^b d(h \circ \gamma) \\
&= \int_a^b \frac{d(h \circ \gamma)(t)}{dt} dt \\
&= h(\gamma(b)) - h(\gamma(a)) \; .
\end{aligned}
$$

The above result is 0 since $\gamma(b) = \gamma(a)$.

The form ω^1 does not satisfy the criterion of Poincaré's Lemma simply because the origin is the singularity of the function and hence is outside its domain of definition. As a matter of fact, ω^1 would not satisfy the theorem of Stokes used above when $M = B^2$ because of this one singularity. These results can be verified using the unit circle centered at the origin S^1 as the reference manifold.

Based on the above observation, we can infer that any extrinsic extension of the element of arclength ds on S^1 or any other differentiable and orientable simple closed curve surrounding the origin cannot be exact. This is because the circumference of such a curve, which is given by the integral of the element of arclength over it, is not zero.

The winding form has higher dimensional analogues in $\mathbb{R}^n$ for $n \geq 3$ as a $(n - 1)$-form on the space. Denoting the winding form on $\mathbb{R}^n \backslash \{0\}$ by ω^{n-1} in each case, we realize the identity:

$$\omega^{n-1}|_{S^{n-1}}(p) = dS_p$$

where the unit sphere S^{n-1} is an embedded hypersurface centered at the origin. Hence, up to sign, integrating the hypersurface element (dS) over S^{n-1} gives the same result as integrating ω^{n-1} over the hypersurface. However, the hypersurface element is always intrinsic to S^{n-1} while ω^{n-1} is extrinsic, and

14

expectedly, the above identity fails to hold when the unit sphere is no longer centered at the origin. When S^{n-1} centered at the origin is displaced by a constant vector $v \in \mathbb{R}^n$ with $\|v\| > 1$, integrating ω^{n-1} over the hypersurface yields 0.

In this case, we can use ω^{n-1} as an extrinsic extension of the hypersurface element (or volume element) of the unit sphere (dS) to the open ambient submanifold $\mathbb{R}^n \backslash \{0\}$ of $\mathbb{R}^n$, when appropriate. We must observe that ω^{n-1} is closed unlike the above mentioned extension

$$\overline{dS} = \sum_{1 \leq i \leq n} N^i (-1)^{i+1} dy_1 \wedge dy_2 \wedge \cdots \wedge \underline{dy_i} \wedge \cdots \wedge dy_n \ ,$$

of dS because ω^{n-1} has a denominator factor. Like ω^1, its denominator causes it to have the origin as a singularity, so it is never exact. Generally speaking, we have the formula

$$\omega^{n-1} = \frac{\sum\limits_{1 \leq i \leq n} y_i (-1)^{i+1} dy_1 \wedge dy_2 \wedge \cdots \wedge \underline{dy_i} \wedge \cdots \wedge dy_n}{\left(\sum\limits_{1 \leq i \leq n} y_i^2 \right)^{\frac{n}{2}}}$$

but the denominator factor is 1 when evaluated on the unit sphere. When working on the sphere or a hypersurface diffeomorphic to the sphere, either extrinsic extension may be taken, depending on the problem or study being addressed.

CHAPTER 2

RIEMANNIAN GEODESICS - AN ILLUSTRATION
FROM THE CALCULUS OF VARIATIONS

Preview. This chapter sheds light on certain essential characteristics of geodesics, which frequently occur in considerations from motion in Euclidean space. Focus is mainly on a method of obtaining them from the calculus of variations, and an explicit geodesic computation for a Riemannian hypersurface.

2.1 Introduction.

Throughout, we will consider smooth, closed hypersurfaces embedded in real Euclidean space. In particular, for non-flat hypersurfaces with non-zero Gaussian curvature, it is evident that straight line segments cannot be admitted within the restrictions of their intrinsic structure. Each geodesic curve must then be identified by way of anayltic devices.

Let E be a real n-dimensional Euclidean space $(n \geq 3)$ and $M \subset E$ be an $(n-1)$-dimensional smooth connected submanifold, in other words, a hypersurface in E. Let $c \subset M$ be a curve admitted by M and parametrized by arclength or a constant multiple of it. Then c is a geodesic of M if and only if $\nabla_{\dot{c}}\dot{c} \equiv 0$, where the velocity vector field $\dot{c}$ along c is simply the derivative of c with respect to the parameter, and ∇ is the covariant derivative. The intrinsic differential operator ∇ is given by the action;

$$\nabla_X Y = D_X Y - \langle D_X Y, \nu \rangle \nu$$

where X and Y are tangential vector fields to the hypersurface, D is the directional derivative operator and ν is the outward unit normal vector field to M.

Clearly, $\dot{c}$ is a tangential vector field to M and

$$\nabla_{\dot{c}} \dot{c} \equiv 0 \quad \Leftrightarrow \quad D_{\dot{c}} \dot{c} \equiv \langle D_{\dot{c}} \dot{c}, \nu \rangle \nu$$
$$\Leftrightarrow \quad D_{\dot{c}} \dot{c} \equiv \| D_{\dot{c}} \dot{c} \| \nu$$

Observe that c is also a submanifold with $\dot{c}$ as the spanning tangential basis vector field. As such, $D_{\dot{c}} \dot{c} = \ddot{c}$ and the above equivalence leads us to

$$\nu = \frac{\ddot{c}}{\|\ddot{c}\|} \quad \text{whenever} \quad \ddot{c} \neq 0.$$

We have $\dot{c}$ in the direction of the principal unit tangent T to c and the principal unit normal N to c is obtained by the formula

$$N = \frac{T'(s)}{\|T'(s)\|}$$

where s is the arclength parameter. But, recalling our chosen parameter,

$$\ddot{c}(ks) = k^2 \ddot{c}(s) = k^2 T'(s)$$

for any constant k, so that

$$\frac{\ddot{c}}{\|\ddot{c}\|} = N.$$

As such, the outward unit normal ν to M and the principal unit normal N to a geodesic $c \subset M$ coincide at each point of c. We take this as an equivalent of the earlier stated geodesic characterization $\nabla_{\dot{c}} \dot{c} \equiv 0$.

With the analogy to motion of rigid bodies along M, taking the time parameter as a constant multiple of the arclength parameter, we get that for a geodesic $c \subset M$, either the acceleration vector $\ddot{c}$ is null or the tangential component to M is null, since ν is parallel to $\ddot{c}$. Of course, $\ddot{c}$ can only be null in the event of motion along a straight line segment.

2.2 Obtaining Geodesic Solutions.

We thus far have the means of identifying the geodesics of hypersurfaces. We now discuss a method of computing the geodesics for a given hypersurface. Typically, we have two settings for such computations, namely;

i.) compute a geodesic curve connecting two given points on the hypersurface,

ii.) compute a geodesic curve given a starting point and tangential direction on the hypersurface.

Although we are guaranteed the existence of solutions in either case due to the Hopf-Rinow theorem, in the first setting we typically do not have uniqueness of solution. Let p and q be two given points on M, so that we denote by $\Omega(M; p, q)$ the set of all smooth curves on M from p to q. This set, referred to as the path space of M from p to q, is itself an infinite-dimensional manifold and this is where all relevant optimization techniques are initiated. This is to say the geodesic computation formulas are derived by differentiation in infinite dimensional (Banach) path spaces. The curves in $\Omega(M; p, q)$ are regarded as points of the infinite-dimensional manifold.

Let $\Gamma \in \Omega(M; p, q)$ be such that $L(\Gamma) = \min\limits_{c \in \Omega(M; p, q)} L(c)$, where $L(c)$ is the length of the curve c. It is well-known that Γ is a geodesic of M but other possible curves in the path space $\Omega(M; p, q)$ which are of greater length and still satisfy the criteria of being smooth geodesic curves are local minimizers of the arclength function in $\Omega(M; p, q)$. This is to say that for each geodesic $\Gamma' \in \Omega(M; p, q)$ there exists a relatively open set U in $\Omega(M; p, q)$ containing Γ' for which $L(\Gamma') = \min\limits_{c \in U} L(c)$. When the geodesic solution is unique, then it gives us the minimizer of the arclength function within the path space being considered. For some clarity, an illustration of the path space for the second setting is hereby provided. Let

$$S := \{\varphi(I) \subseteq \mathbb{R}^n : \varphi \in C^2(\overline{I}, \mathbb{R}^n)\}$$

and define

$$\varphi_1(I) + \varphi_2(I) := \{\varphi_1(t) + \varphi_2(t) : t \in I\}$$

for a bounded subinterval I of the reals, so that S is a linear space. As a hypersurface, M may be given by the formula $g_M{}^{-1}\{k\}$ where $k \in \mathbb{R}$ is constant and g_M is a smooth functional on $\mathbb{R}^n$. In this event, the path space $\Omega(M; p, \mathbf{v}) \subset S$ equals the set

$$\left\{\varphi(I) \in S : \varphi(\inf I) = p \;;\; \frac{d\varphi}{dt}\Big|_{\inf I} = \left\|\frac{d\varphi}{dt}\Big|_{\inf I}\right\| . \mathbf{v} \;;\; g_M(\varphi(t)) = k \; \forall t \in I\right\}$$

and this is precisely the set of all smooth curves on M starting from $p \in M$ in the unit direction $\mathbf{v}_p$ from the tangent space T_pM. $\Omega(M; p, \mathbf{v})$ is also infinite dimensional, but each geodesic in $\Omega(M; p, \mathbf{v})$ is either an extension or restriction of a unique $\Gamma \in \Omega(M; p, \mathbf{v})$. Moreover, if M can be represented by a periodic parametrization then we have a unique periodic extension for Γ on M. We denote lengths of curves $c \in \Omega(M; p, \mathbf{v})$ by $L_{I,t}(c)$ where t is a particular parameter running through the interval I.

Several constants of integration emerge in the course of geodesic computations, and because of guaranteed uniqueness of these constants with respect to a given co-ordinate system for the second case $\Omega(M; p, \mathbf{v})$, this setting is better amenable than the previously discussed path space $\Omega(M; p, q)$.

The arclength function in the n-dimensional space E is given by the indefinite integral

$$\int \sqrt{\sum_{i=1}^{n} \pi_i{}^2} \,,$$

where $\{\pi_i\}_{i=1}^{n}$ is the set of canonical projection maps on E. Taking E to be $\mathbb{R}^n$ with the Euclidean co-ordinate system $(x_1, x_2, \cdots, x_n)$ then the projection maps π_i are written as dx_i for $i = 1, 2, \cdots, n$ and consequently, the arclength function is given by

$$s = \int \sqrt{\sum_{i=1}^{n} dx_i{}^2} \,.$$

Expressing each x_i in terms of a common parameter t running through an interval $I \subseteq \mathbb{R}$ (which we can always do for a smooth curve in Euclidean space), we get the arclength function $:= s_{I,t}$ as a definite integral

$$s_{I,t} = \int_I \sqrt{\sum_{i=1}^{n} \left(\frac{dx_i}{dt}\right)^2} \, dt \,.$$

We have $s_{I,t}$ as a functional map on the path space $\Omega(M; p, \mathbf{v})$;

$$s_{I,t} : \quad \Omega(M; p, \mathbf{v}) \quad \to \quad \mathbb{R}$$
$$c \quad \mapsto \quad L_{I,t}(c),$$

where $L_{I,t}(c)$ retains its previous notation.

The arclength function $s_{I,t}$ on the considered path spaces is convex and continuously differentiable, but the lack of reflexivity of the path spaces thwarts arguments about uniqueness of geodesic solutions, in consideration of classical optimization theorems of functional analysis [2] (p. 155). However, in light of the uniqueness of the geodesic (and thus also the minimizer) in our current setting, any relevant optimization device employed here will yield the unique solution.

For instance, the canonical Euler - Lagrange equations are derived by differentiation in the infinite-dimensional path spaces, and they are popularly used for this type of problem. We have a functional of the type

$$s_{I,t} = \int_I \Lambda(t, u_1(t), u_2(t), \cdots, u_{n-1}(t), \dot{u}_1, \dot{u}_2, \cdots, \dot{u}_{n-1}) dt$$

taking M to be parametrized by the co-ordinate system

$$f : U \subseteq \mathbb{R}^{n-1} \to M \subset \mathbb{R}^n \; ; \; (u_1, u_2, \cdots, u_{n-1}) \mapsto (x_1, x_2, \cdots, x_n) \; ,$$

where $U \subseteq \mathbb{R}^{n-1}$ is a cube, or a convex domain for which $\overset{\circ}{U} \neq \emptyset$. By the canonical Euler-Lagrange equations, all minimizing points of $s_{I,t}$ must satisfy

$$\Lambda_{u_i} = \frac{d}{dt}(\Lambda_{\dot{u}_i})$$

for $1 \leq i \leq n - 1$. Here, $\dot{u}_i$ denotes $\dfrac{du_i}{dt}$ and Y_x denotes partial derivation of Y with respect to the variable connoted by the subscript x. We obtain the transformation of $s_{I,t}$ from its initial form $\displaystyle\int_I \sqrt{\sum_{i=1}^{n} \left(\frac{dx_i}{dt}\right)^2} \, dt$ by way of the Riemannian structure $\{g_{ij}\}_{i,j=1}^{n-1}$ obtained from the co-ordinate system f where $g_{ij} = \langle \dfrac{\partial f}{\partial u_i}, \dfrac{\partial f}{\partial u_j} \rangle$. The function $s_{I,t}$ then becomes

$$\int_I \sqrt{\sum_{i,j=1}^{n-1} g_{ij} \dot{u}_i \dot{u}_j} \, dt$$

which is the suitable form to apply the Euler - Lagrange equations.

We can derive the weak formulation for the Euler-Lagrange equations by considering path spaces one more time. This time we work in

$$S^* := \{\varphi(I) \subseteq \mathbb{R}^{n-1} : \varphi \in C^1(\overline{I}, \mathbb{R}^{n-1})\}$$

20

with the norm

$$||\varphi|| = ||\varphi||_\infty + ||\dot\varphi||_\infty$$

where

$$||x||_\infty = \sup_{t\in\bar{I}}||x(t)||_2$$

giving us a Banach structure on $(S^*, ||.||)$. Let $p = f(a)$ and $Df_p(\psi) := f'(a; \psi) = \mathbf{v}$. Hence, we are minimizing the functional $s_{I,t}$ over the convex subset $\Omega(f^{-1}(M); a, \psi) \subseteq S^*$ which is an important advantage of computing in this setting instead of in the non-convex $\Omega(M, p, \mathbf{v}) \subset S$. Assuming that $\overline{\varphi}$ is a local minimizer of $s_{I,t}$ in $\Omega(f^{-1}(M); a, \psi) \subseteq S^*$ then

$$\exists r > 0 : s_{I,t}(\overline{\varphi}) \le s_{I,t}(\varphi) \ \forall \varphi \in \Omega(f^{-1}(M); a, \psi) \cap B(\overline{\varphi}, r) \ .$$

In this setting, for any $\zeta \in S^*$ satisfying $\zeta(\inf I) = \zeta(\sup I) = \frac{d\zeta}{dt}|_{\inf I} = 0$ we have $\overline{\varphi} + \tau\zeta \in \Omega(f^{-1}(M); a, \psi)$ where τ runs through an open interval containing zero. Also, there exists some real number $\delta > 0$ such that for all $\tau \in (-\delta, \delta)$, we have $\overline{\varphi} + \tau\zeta \in B(\overline{\varphi}, r)$. Define $\gamma(\tau) := s_{I,t}(\overline{\varphi} + \tau\zeta)$ so that $\gamma(0) \le \gamma(\tau) \ \forall \tau \in (-\delta, \delta)$. In other words, 0 is a minimizer of γ in $(-\delta, \delta)$ which means $\gamma'(0) = 0 \ \Rightarrow \ s'_{I,t}(\overline{\varphi}).\zeta = 0$.

$$
\begin{aligned}
s'_{I,t}(\overline{\varphi}).\zeta &= \lim_{\alpha\to 0}\left(\frac{s_{I,t}(\overline{\varphi} + \alpha\zeta) - s_{I,t}(\overline{\varphi})}{\alpha}\right) \\
&= \int_I \lim_{\alpha\to 0}\frac{\Lambda(t,\overline{\varphi}(t) + \alpha\zeta(t), \dot{\overline{\varphi}}(t) + \alpha\dot\zeta(t)) - \Lambda(t,\overline{\varphi}(t), \dot{\overline{\varphi}}(t))}{\alpha}dt.
\end{aligned}
$$

The quantity under the above integral is identified as the following directional derivative in $\mathbb{R} \times \mathbb{R}^{n-1} \times \mathbb{R}^{n-1}$;

$\Lambda'(t,\overline{\varphi}(t), \dot{\overline{\varphi}}(t)); (0, \zeta(t), \dot\zeta(t))$
$= \Lambda_u(t,\overline{\varphi}(t), \dot{\overline{\varphi}}(t)).\zeta(t) + \Lambda_{\dot u}(t,\overline{\varphi}(t), \dot{\overline{\varphi}}(t)).\dot\zeta(t)$

where $u = (u_1, u_2, \cdots, u_{n-1})$ and $\dot u = (\dot u_1, \dot u_2, \cdots, \dot u_{n-1})$. We justify passing the limit into the above integral by way of uniform convergence of the integrand at differentiable points of Λ in the limit as $\alpha \to 0$. Resultantly, the weak formulation of the Euler-Lagrange equations is

$$\int_I \left(\Lambda_u(t,\overline{\varphi}(t), \dot{\overline{\varphi}}(t)).\zeta(t) + \Lambda_{\dot u}(t,\overline{\varphi}(t), \dot{\overline{\varphi}}(t)).\dot\zeta(t)\right) dt = 0$$

21

for all $\zeta \in S^*$ satisfying $\zeta(inf I) = \zeta(sup I) = \frac{d\zeta}{dt}|_{inf I} = 0$. Integrating by parts and assuming in addition that $\overline{\varphi}$ is of class C^2, we have

$$\int_I \left(\Lambda_u(t, \overline{\varphi}(t), \dot{\overline{\varphi}}(t)) - \frac{d}{dt}\Lambda_{\dot{u}}(t, \overline{\varphi}(t), \dot{\overline{\varphi}}(t)) \right) . \zeta dt = 0 \ .$$

By the basic lemma of the calculus of variations, the first vector in the integrand; $\Lambda_u(t, \overline{\varphi}(t), \dot{\overline{\varphi}}(t)) - \frac{d}{dt}\Lambda_{\dot{u}}(t, \overline{\varphi}(t), \dot{\overline{\varphi}}(t)) \in \mathbb{R}^{n-1}$ vanishes giving us

$$\Lambda_{u_i} = \frac{d}{dt}(\Lambda_{\dot{u}_i})$$

for $1 \leq i \leq n-1$. Hence the computed curve lies in $\mathbb{R}^{n-1}$ while the geodesic is its image under f.

Note that geodesics are intrinsic to their manifolds, and so are invariant under which co-ordinate systems are used to derive them. Nevertheless, a suitable choice of co-ordinate system could significantly ease a computation. Different manifolds and their co-ordinate systems present their own challenges but the main theoretical properties are tantamount in all applications.

In motion along manifolds, minimal geodesics are the 'lazy' curves which minimize both the arclength function and energy functionals. For the path space $\Omega(M; p, \mathbf{v})$, a kinetic energy functional is given by

$$U_{I,t} : \quad \Omega(M; p, \mathbf{v}) \quad \rightarrow \quad \mathbb{R}$$
$$c \quad \mapsto \quad \int_I \|\dot{c}(t)\|^2 dt$$

and another is given by

$$c \longmapsto \int_I \kappa_c^2 ds \ ,$$

where κ_c is the curvature of c in terms of the arclength parameter.

In essence, for a particle at rest on a smooth Riemannian hypersurface, when a linear tangential force is applied to it so that it undergoes motion on the hypersurface without slipping, it is translated along the geodesic path. In vector calculus, the phenomenon of geodesics is central to several themes, including parallel transports and Jacobi fields.

2.3 Computation of Geodesic Equation for Unit 3-Sphere in $\mathbb{R}^4$.

With all the given background theoretical information, we now proceed to demonstrate the computation of a hyperspherical geodesic. According to a concise argument by Jost and Li-Jost[5] (p. 42), all such curves must be portions of great circles of the hypersphere. We will operate using the hyperspherical co-ordinate system, which yields geodesic equations easier to solve in comparison to most other known co-ordinate systems. The minimal global parametrization of the unit sphere $S^{n-1} \subset \mathbb{R}^n$ by the hyperspherical co-ordinate system for $n \geq 3$ is given by

$$f_{(n-1)} : [\tfrac{-\pi}{2}, \tfrac{\pi}{2}]^{n-2} \times [0, 2\pi] \longrightarrow S^{n-1} \; ; (u_1, u_2, \cdots, u_{n-1}) \longmapsto (x_1, x_2, \cdots, x_n);$$

$$x_1 = \prod_{j=1}^{n-1} \cos u_j \; ; \quad x_i = \prod_{j=1}^{n-i} \cos u_j \sin u_{n-i+1} \quad [2 \leq i \leq n-1],$$
$$x_n = \sin u_1 \; .$$

For the case of interest of least dimension (S^2), we have a multiple parametrization

$$\overline{f_2} : [\tfrac{-\pi}{2}, \tfrac{\pi}{2}] \times \mathbb{R} \longrightarrow \mathbb{R}^3;$$
$$\overline{f_2}^1 (u_1, u_2) := x = \cos u_1 \cos u_2$$
$$\overline{f_2}^2 (u_1, u_2) := y = \cos u_1 \sin u_2$$
$$\overline{f_2}^3 (u_1, u_2) := z = \sin u_1 \; .$$

Pre-images of geodesic solutions computed in $[\tfrac{-\pi}{2}, \tfrac{\pi}{2}] \times \mathbb{R}$ are known to have

a parametrization $\left(u_1, \arctan\left[\dfrac{\sin u_1}{\sqrt{\dfrac{\cos^2 u_1}{\gamma^2} - 1}} \right] + \delta \right)$ for real constants of

integration δ and $\gamma, |\gamma| < 1$. This solution holds only for $|u_1| < \arccos(|\gamma|)$. Singularities occur at $u_1 = \pm \arccos(|\gamma|)$, where we respectively take $u_2 = \pm\tfrac{\pi}{2} + \delta$ and for $|u_1| > \arccos(|\gamma|)$, we have no real solutions. This is due to the fact that the solution curve does not cut across the parallels $u_1{}^*$ of S^2

for which $|u_1{}^*| > \arccos(|\gamma|)$. We extend the solution into another branch

$$\left\{\left(u_1, \pi - \arctan\left[\frac{\sin u_1}{\sqrt{\dfrac{\cos^2 u_1}{\gamma^2} - 1}}\right] + \delta\right)\right\}_{|u_1| < \arccos(|\gamma|)}$$

to get one full cycle of the geodesic curve, considering the minimal domain of $\overline{f_2}$. For either branch, the image computed under $\overline{f_2}$ yields one half of a great circle of (S^2). The integration constant γ equals 0 if and only if the solution curve is a meridian and the special case of $|\gamma| = 1$ occurs if and only if the solution is the equator.

We will now proceed with the computation of a geodesic equation for the unit hypersphere $S^3 \subset \mathbb{R}^4$. It is given once by the global co-ordinate system,

$$f_3: \quad [\tfrac{-\pi}{2}, \tfrac{\pi}{2}]^2 \times [0, 2\pi] \quad \longrightarrow \quad S^3 \subset \mathbb{R}^4$$
$$(u_1, u_2, u_3) \quad \longmapsto \quad f(u_1, u_2, u_3) = (x_1, x_2, x_3, x_4)$$

$$f_3(u_1, u_2, u_3) = (\cos u_1 \cos u_2 \cos u_3, \cos u_1 \cos u_2 \sin u_3, \cos u_1 \sin u_2, \sin u_1)\,.$$

S^3 is also covered multiple times by extending f_3 to $\overline{f_3} := f$ with a domain $[\tfrac{-\pi}{2}, \tfrac{\pi}{2}] \times \mathbb{R}^2$ and the same parametrization in the co-domain $\mathbb{R}^4$. In both cases, using the notation $g_{ij} = \langle \dfrac{\partial f}{\partial u_i}, \dfrac{\partial f}{\partial u_j}\rangle$, we get

$g_{11} = 1$
$g_{22} = \cos^2 u_1$
$g_{33} = \cos^2 u_1 . \cos^2 u_2$
$g_{ij} = 0, \quad$ for $i \neq j$.

Setting the common parameter to be u_1, we get an arclength function on the path space $\Omega(S^3; p, \mathbf{v})$ for $\mathbf{v}_p \in T_p S^3$;

$$s_{I,u_1} = \int_I \sqrt{1 + \cos^2 u_1 \left(\frac{du_2}{du_1}\right)^2 + \cos^2 u_1 \cos^2 u_2 \left(\frac{du_3}{du_1}\right)^2}\, du_1 := \int_I \Lambda du_1$$

for an appropriate interval $I \subseteq [\tfrac{-\pi}{2}, \tfrac{\pi}{2}]$. In determining the minimizer c of s_{I,u_1}, our choice of parameter ideally ensures that $\|c'(u_1)\| \neq 0$ on I and it leaves us to set up the following Euler - Lagrange equations (1) and (2)- -

For the first Euler - Lagrange equation,

$$\Lambda_{u_3} = \frac{d}{du_1}(\Lambda_{\dot{u}_3})$$

$$\Leftrightarrow \left(\frac{du_3}{du_1}\right)^2 = \frac{\gamma^2\left(\sec^2 u_1 + (\frac{du_2}{du_1})^2\right)}{\cos^2 u_1 \cos^4 u_2 - \gamma^2 \cos^2 u_2} \quad \cdots (1)$$

for a real constant $\gamma, |\gamma| < 1$.

For the second Euler - Lagrange equation,

$$\Lambda_{u_2} = \frac{d}{du_1}(\Lambda_{\dot{u}_2})$$

$$\Leftrightarrow \frac{-1}{\Lambda}(\cos^2 u_1 \cos u_2 \sin u_2)(\dot{u}_3)^2 = \frac{d}{du_1}\left(\frac{\dot{u}_2 \cos^2 u_1}{\Lambda}\right)$$

$$= \frac{\Lambda(-2\dot{u}_2 \cos u_1 \sin u_1 + \ddot{u}_2 \cos^2 u_1) - \dot{u}_2(\frac{d\Lambda}{du_1})\cos^2 u_1}{\Lambda^2} \quad \cdots (2).$$

Consistent with previous notation, $\dot{Y} = \dfrac{dY}{du_1}$ and $\ddot{Y} = \dfrac{d^2Y}{du_1^2}$.

After grinding out with lengthy arithmetic and substituting (1) in (2), we are able to obtain a differential equation in exactly 2 variables:

$$\dot{u}_2 \sin u_1 \cos u_2(\gamma^2 - 2\cos^2 u_1 \cos^2 u_2) + \ddot{u}_2 \cos u_1 \cos u_2(\cos^2 u_1 \cos^2 u_2 - \gamma^2) +$$
$$\gamma^2 \sec u_1 \sin u_2 + \gamma^2(\dot{u}_2)^2 \cos u_1 \sin u_2 - (\dot{u}_2)^3 \cos^4 u_1 \sin u_1 \cos^3 u_2 = 0 \quad \cdots (3)$$

It is useful to say how to obtain γ by way of simple vector calculus. Already having $\mathbf{v}_p \in T_pS^3$ and the gradient of f at p, Df_p, we obtain the pre-image of the position vector $\mathbf{v}_p$ under Df which we denote ψ so that $Df_p(\psi) = \mathbf{v}_p$. Now, having the vector $\psi := (\psi_1, \psi_2, \psi_3)$ and the pre-image of the point p under f, we substitute $du_3(\psi) = \psi_3, du_2(\psi) = \psi_2$ and $du_1(\psi) = \psi_1$ in (1) to realize the value of γ. ($\{du_i\}_{i=1}^{3}$ is the set of canonical projection maps on the domain $\mathbb{R}^3$.)

Upon inspection, $u_2 \equiv 0$ is a solution of the non-linear differential equation (3). This solution occurs if and only if the u_2 co-ordinate at the initial position is 0 and the pullback of the initial velocity vector $\mathbf{v}_p$ has a null second component, which is to say $\psi_2 = 0$. To find the general solution for (3), we re-arrange the equation as follows –

$[1 + (\frac{du_2}{du_1})^2 \cos^2 u_1] \, [\gamma^2 \sec^2 u_1 \tan u_2 - (\frac{du_2}{du_1}) \tan u_1 \cos^2 u_1 \cos^2 u_2] =$
$[\gamma^2 - \cos^2 u_1 \cos^2 u_2] \, [\frac{d^2 u_2}{du_1^2} - (\frac{du_2}{du_1}) \tan u_1]$

The equation then becomes easier to decompose,

$$\frac{\ddot{u}_2 - \dot{u}_2 \tan u_1}{1 + (\dot{u}_2)^2 \cos^2 u_1} = \frac{\gamma^2 \sec^2 u_1 \tan u_2 - \dot{u}_2 \tan u_1 \cos^2 u_1 \cos^2 u_2}{\gamma^2 - \cos^2 u_1 \cos^2 u_2}$$

$$\Leftrightarrow \quad \frac{\frac{d}{du_1}(1 + (\dot{u}_2)^2 \cos^2 u_1)}{2\dot{u}_2 \cos^2 u_1 [1 + (\dot{u}_2)^2 \cos^2 u_1]} = \frac{\gamma^2 \sec^2 u_1 \tan u_2 - \dot{u}_2 \tan u_1 \cos^2 u_1 \cos^2 u_2}{\gamma^2 - \cos^2 u_1 \cos^2 u_2}$$

$$\Leftrightarrow \quad \frac{\frac{d}{du_1}(1 + (\dot{u}_2)^2 \cos^2 u_1)}{1 + (\dot{u}_2)^2 \cos^2 u_1} = \frac{2\gamma^2 \dot{u}_2 \tan u_2 \sec^2 u_2 - 2(\dot{u}_2)^2 \tan u_1 \cos^4 u_1}{\gamma^2 \sec^2 u_2 - \cos^2 u_1}$$

$$= \frac{2\gamma^2 \dot{u}_2 \tan u_2 \sec^2 u_2 + 2 \cos u_1 \sin u_1 - 2 \cos u_1 \sin u_1 [1 + (\dot{u}_2)^2 \cos^2 u_1]}{\gamma^2 \sec^2 u_2 - \cos^2 u_1}$$

$$= \frac{\frac{d}{du_1}(\gamma^2 \sec^2 u_2 - \cos^2 u_1)}{\gamma^2 \sec^2 u_2 - \cos^2 u_1} - \frac{2 \cos u_1 \sin u_1 [1 + (\dot{u}_2)^2 \cos^2 u_1]}{\gamma^2 \sec^2 u_2 - \cos^2 u_1}$$

Integrating both sides with respect to u_1 , we get

$$ln[1 + (\dot{u}_2)^2 \cos^2 u_1] = ln[\gamma^2 \sec^2 u_2 - \cos^2 u_1] - \int \frac{2 \cos u_1 \sin u_1 [1 + (\dot{u}_2)^2 \cos^2 u_1]}{\gamma^2 \sec^2 u_2 - \cos^2 u_1} du_1$$

$$\Leftrightarrow ln\left(\frac{1 + (\dot{u}_2)^2 \cos^2 u_1}{\gamma^2 \sec^2 u_2 - \cos^2 u_1}\right) = -\int 2 \cos u_1 sin u_1 \left(\frac{1 + (\dot{u}_2)^2 \cos^2 u_1}{\gamma^2 \sec^2 u_2 - \cos^2 u_1}\right) du_1$$

Hence, we solve for U in the equation;

$$\int U(t).V(t) dt = ln(U(t)) \quad [t = u_1]$$

$$\Leftrightarrow \quad U.V = \left(\frac{1}{U}\right)\frac{dU}{dt}$$

$$\Leftrightarrow \quad \frac{\frac{dU}{dt}}{U^2} = V$$

$$\Leftrightarrow \quad -\frac{1}{U} = \int V dt$$

Swapping back U with $\dfrac{1 + (\dot{u}_2)^2 \cos^2 u_1}{\gamma^2 \sec^2 u_2 - \cos^2 u_1}$ and V with $-2\cos u_1 \sin u_1$,

we get (3) as an equivalent separable differential equation - -

$$\frac{\gamma^2 \sec^2 u_2 - \cos^2 u_1}{1 + (\dot{u}_2)^2 \cos^2 u_1} = -\cos^2 u_1 + k$$

$$\Leftrightarrow \qquad \gamma^2 \sec^2 u_2 = -(\dot{u}_2)^2 \cos^4 u_1 + k + k(\dot{u}_2)^2 \cos^2 u_1$$

$$\Leftrightarrow \qquad \gamma^2 \sec^2 u_2 - k = \left(\frac{du_2}{du_1}\right)^2 [k \cos^2 u_1 - \cos^4 u_1] \qquad \cdots (3')$$

The constant of integration k above is obtained in a similar way as γ.

(*)Apologetically, we will fix k here to equal $2\gamma^2 - \gamma^4$ in order to shorten the computation and avoid carrying along too many constants of integration. With this substitution, we have as a non-trivial solution for $(3')$;

$$u_2 = \arctan\left[\frac{\sin u_1}{\sqrt{\dfrac{\cos^2 u_1}{\gamma^2} - 1}}\right] = \arcsin\left[\frac{\tan u_1}{\sqrt{\dfrac{1}{\gamma^2} - 1}}\right] \quad \text{for } |u_1| < \arccos(\sqrt{2\gamma^2 - \gamma^4})$$

$$\Rightarrow \frac{du_2}{du_1} = \frac{\gamma \sec u_1}{\sqrt{\cos^2 u_1 - \gamma^2}} \quad \text{and}$$

$$\cos u_2 = \left(\frac{\dfrac{1}{\gamma^2} - \sec^2 u_1}{\dfrac{1}{\gamma^2} - 1}\right)^{\frac{1}{2}}.$$

Henceforth, we will write t instead of u_1. From equation (1), we have

$$\left(\frac{du_3}{dt}\right)^2 = \left(\sec^2 t + \frac{\gamma^2 \sec^2 t}{\cos^2 t - \gamma^2}\right)\left(\frac{\gamma^2}{\cos^2 t \left(\frac{1 - \gamma^2 \sec^2 t}{1 - \gamma^2}\right)^2 - \frac{\gamma^2 - \gamma^4 \sec^2 t}{1 - \gamma^2}}\right)$$

$$\Rightarrow \left|\frac{du_3}{dt}\right| = \frac{|\gamma|(1 - \gamma^2)}{\sqrt{(\cos t - \gamma^2 \sec t)^2[\cos^2 t - 2\gamma^2 + \gamma^4]}} \qquad \cdots (4)$$

A solution for (4) when $|t| < \arccos(\sqrt{2\gamma^2 - \gamma^4})$ is

$$u_3 = \arcsin\left(\frac{\tan t}{\sqrt{\dfrac{1}{\gamma^2} - 1}\sqrt{1 - \gamma^2 \sec^2 t}}\right) + \beta$$

27

The constant β is determined by obtaining the pre-image of the starting point p under f and substituting in the solution above. The above solution given for u_3 runs through an interval of length π. Given the minimal domain of f, we extend this solution into another branch, namely - -

$$\overline{u_3} = \pi - \arcsin\left(\frac{\tan t}{\sqrt{\frac{1}{\gamma^2} - 1}\sqrt{1 - \gamma^2 \sec^2 t}}\right) + \beta$$

so that the third argument of the domain runs through an interval of length 2π, hence completing one full cycle of the solution curve. Let us set $z = u_3 - \beta$ (resp. $z = \overline{u_3} - \beta$) so that we have;

$$\cos t \cos u_2 \cos u_3 = \cos t \cos u_2 (\cos z \cos \beta - \sin z \sin \beta)$$

$$= \pm \cos t \left(\frac{\frac{1}{\gamma^2} - \sec^2 t}{\frac{1}{\gamma^2} - 1}\right)^{\frac{1}{2}} \left(\frac{\frac{1}{\gamma^2} + (\gamma^2 - 2)\sec^2 t}{(\frac{1}{\gamma^2} - 1)(1 - \gamma^2 \sec^2 t)}\right)^{\frac{1}{2}} \cos \beta -$$

$$\cos t \left(\frac{\frac{1}{\gamma^2} - \sec^2 t}{\frac{1}{\gamma^2} - 1}\right)^{\frac{1}{2}} (\sin z)(\sin \beta)$$

$$= \frac{\pm \cos t \left(\frac{1}{\gamma^2} + (\gamma^2 - 2)\sec^2 t\right)^{\frac{1}{2}} \cos \beta}{\gamma(\frac{1}{\gamma^2} - 1)} - \frac{\sin t \sin \beta}{\gamma(\frac{1}{\gamma^2} - 1)}$$

$$\cos t \cos u_2 \sin u_3 = \cos t \cos u_2 (\cos z \sin \beta + \sin z \cos \beta)$$

$$= \frac{\pm \cos t \left(\frac{1}{\gamma^2} + (\gamma^2 - 2)\sec^2 t\right)^{\frac{1}{2}} \sin \beta}{\gamma(\frac{1}{\gamma^2} - 1)} + \frac{\sin t \cos \beta}{\gamma(\frac{1}{\gamma^2} - 1)}$$

$$\cos t \sin u_2 = \frac{\sin t}{\sqrt{\frac{1}{\gamma^2} - 1}}$$

Setting the parameters $\varphi_1|_{S^3} = \sin t$ and $\varphi_2|_{S^3} = \cos t \left(\frac{1}{\gamma^2} + (\gamma^2 - 2)\sec^2 t\right)^{\frac{1}{2}}$, we see that the solution curve lies on the 2-plane in $\mathbb{R}^4$ parametrized by

$$\left(\frac{(\cos \beta)\varphi_2 - (\sin \beta)\varphi_1}{\gamma(\frac{1}{\gamma^2} - 1)}, \; \frac{(\sin \beta)\varphi_2 + (\cos \beta)\varphi_1}{\gamma(\frac{1}{\gamma^2} - 1)}, \; \frac{\varphi_1}{\sqrt{\frac{1}{\gamma^2} - 1}}, \; \varphi_1\right).$$

Since this plane passes through the origin in $\mathbb{R}^4$, its intersection with S^3, which is precisely the locus of our solution curve, is a great circle of the hypersphere as expected. The curve does not intersect sections of the hypersurface for which $|u_1| > \arccos(\sqrt{2\gamma^2 - \gamma^4})$ and it is left as an exercise for the reader to verify that

$$L_{I,u_1}(c) := \int_{|u_1|<\arccos(\sqrt{2\gamma^2-\gamma^4})} \|Df(\dot{c})\| du_1 = \pi$$

holds for either solution branch c in the domain of f.

The graph below depicts the pre-image of our solution curve under f in $\mathbb{R}^3$ for $\beta = 0$ and $\gamma = \frac{\sqrt{2}}{2}$.

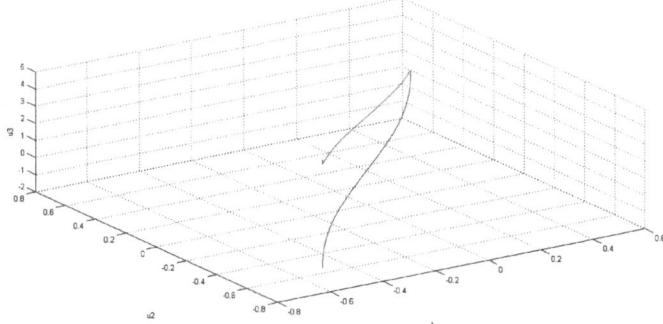

Due to the action taken earlier in this section (*) dropping another constant of integration, the solution we have corresponds to only a single integral curve of equation (3′). All geodesics of S^3 correspond to the integral curves of (3′), which depend on the constant γ to be determined beforehand. With these observations, we have the extrapolation of the 2-spherical geodesic equation into one higher dimension. Although somewhat tedious in leaping from the second dimension which is of greatest interest, this should not be regarded as a foolhardy venture as all results are obtainable in terms of elementary functions and hold potential for beneficial analysis.

CHAPTER 3

SOLVING DIFFERENTIAL EQUATIONS BY SYMMETRY GROUPS

Preview. In light of the ever-growing relevance of obtaining precise analytic solutions to differential equations in numerous fields of mathematics and other computational sciences, the method of symmetry groups is a device that can effectively be used to determine certain solutions called invariant solutions. This device can prove especially useful for non-linear models of ordinary first-order equations which are inseparable, higher order Ordinary Differential Equations (O.D.E's) and Partial Differential Equations (P.D.E's). When a differential equation is observed to admit a symmetry Lie group of transformations, it significantly eases the state of the problem at hand. The aim of this chapter is to describe the critical properties of Lie groups and then introduce their role in the simplification process of differential equations with illustrations. There is particular emphasis on the admittance of the special group SL(3, $\mathbb{R}$) by all second order linear O.D.E's by way of a Kummer-Liouville transformation.

3.1 Lie Groups.

The underlying concept of a Lie group is quite simple. It is a differentiable manifold G endowed with an abstract group structure such that the map

$G \times G \to G$; $(\sigma, \tau) \mapsto \sigma\tau^{-1}$ is C^{∞}. The most important category of Lie groups is made of the matrix groups, which are all subgroups of the spaces of invertible matrices $\mathrm{GL}(n, \mathbb{R})$ for $n \geq 2$. We observe first that

$$GL(n, \mathbb{R}) = \{A \in \mathcal{M}_{n \times n}(\mathbb{R}) : Det(A) \neq 0\}$$

is an open subset of the space of square matrices, and is therefore a submanifold. A popular result of Carter and Von Neumann asserts that each closed subgroup of $\mathrm{GL}(n, \mathbb{R})$ is also a manifold and therefore a Lie group. Notably, each linear Lie group is isometrically isomorphic to one of the matrix groups. Examples of linear Lie groups are as follows;
i.) the special Lie group $SL(n, \mathbb{R})$ which is the set of matrices in $\mathrm{GL}(n, \mathbb{R})$ with determinant equal to 1 ,
ii.) the group of orthogonal matrices $O(n)$ which is the set of $n \times n$ matrices A such that $A^T = A^{-1}$,
iii.) the group of rotation matrices $SO(n) := SL(n, \mathbb{R}) \cap O(n)$.

It is easy to check that each of these form non-abelian groups with the group operation of matrix multiplication. Not to be separated from the mention of a Lie group is its Lie algebra, which is defined to be the tangent space to the Lie group at the identity. The Lie algebra of $\mathrm{GL}(n, \mathbb{R})$ is $\mathcal{M}_{n \times n}(\mathbb{R})$ since the topological subspace of invertible matrices is an open subset of the collection of square matrices. As further examples, the Lie algebras of $SL(n, \mathbb{R}), O(n), SO(n)$ are respectively denoted $sl(n, \mathbb{R}), o(n), so(n)$ and they are real vector spaces whose dimensions coincide with those of the corresponding Lie groups. $sl(n, \mathbb{R})$ is the set of real $n \times n$ matrices with null trace, while $o(n) = so(n)$ is the set of real skew-symmetric $n \times n$ matrices. The Lie algebra is often described to be a linearization of the Lie group near the identity element. The exponential map exp provides a means of returning from the Lie algebra to the Lie group by a process called delinearization of the Lie algebra. Although the exponential map is generally not surjective, it is important to note that every one parameter subgroup of a given Lie group is of the form $exp(\lambda A)$ for some A in the Lie algebra.

3.1.1 One-Parameter Lie Groups

We will begin with the fundamental setting of Lie groups acting on a plane. Let $\mathbf{x} = (x, y)$ and $\mathbf{X} = (X, Y)$ be points in $\mathbb{R}^2$ and for $\lambda \in \mathbb{R}$, let $P_\lambda : \mathbf{x} \mapsto f(\mathbf{x}, \lambda) = \mathbf{X}$ be a transformation depending on the parameter λ

mapping points $\mathbf{x}$ to $\mathbf{X}$. Assigning an additive composition to the group of λ, that is, $\psi(\lambda_1, \lambda_2) = \lambda_1 + \lambda_2$, we say that the set of transformations P_λ is an additive transformation group if the following four conditions are satisfied:

1. P_λ is one - to - one.

2. $P_{\lambda_2} \circ P_{\lambda_1} = P_{\lambda_2 + \lambda_1}$, that is, $f(f(\mathbf{x}, \lambda_1), \lambda_2) = f(\mathbf{x}, \lambda_2 + \lambda_1)$

3. $P_0 = Identity$, that is, $f(\mathbf{x}, 0) = \mathbf{x}$

4. $P_\lambda \circ P_{-\lambda} = Identity$, where for each λ, $-\lambda$ is uniquely determined.

If, in addition, f is infinitely differentiable with respect to $\mathbf{x}$ and λ, we say that the group $\{P_\lambda\}$ is a one-parameter Lie group of transformations. A few common one parameter Lie groups of transformations are identified as follows in their global form;

(i.) The translational group or trivial group: $X = x$, $Y = y + \lambda$
(ii.) The stretching group: $X = e^\lambda x$, $Y = e^\lambda y$
(iii.) The rotation group: $X = x cos\lambda - y sin\lambda$, $Y = x sin\lambda + y cos\lambda$.

Observe in the case of the stretching group, we may express the linear transformation as

$$\begin{bmatrix} X \\ Y \end{bmatrix} = \begin{bmatrix} e^\lambda & 0 \\ 0 & e^\lambda \end{bmatrix} \cdot \begin{bmatrix} x \\ y \end{bmatrix}.$$

Likewise in the case of the rotation group, we may express the transformation as

$$\begin{bmatrix} X \\ Y \end{bmatrix} = \begin{bmatrix} cos\lambda & -sin\lambda \\ sin\lambda & cos\lambda \end{bmatrix} \cdot \begin{bmatrix} x \\ y \end{bmatrix}.$$

It can easily be shown that

$$exp\left(\lambda \cdot \begin{bmatrix} 1 & 0 \\ 0 & 1 \end{bmatrix}\right) = \begin{bmatrix} e^\lambda & 0 \\ 0 & e^\lambda \end{bmatrix}$$

and

$$exp\left(\lambda \cdot \begin{bmatrix} 0 & 1 \\ -1 & 0 \end{bmatrix}\right) = \begin{bmatrix} cos\lambda & -sin\lambda \\ sin\lambda & cos\lambda \end{bmatrix}.$$

As such, we can draw certain useful inferences by way of the exponential map, recalling that each one parameter subgroup of a given Lie group is of the form $exp(\lambda A)$ for some A in the Lie algebra. For instance, we can infer that the rotation group $SO(2)$ possesses the Lie algebra which is the set of 2×2 skew-symmetric matrices, denoted $so(2)$. More broadly, we can infer also that the rotation group is a one parameter subgroup of the matrix group whose Lie algebra is the set of matrices with zero trace, which is denoted $sl(2, \mathbb{R})$. Of course, this corresponds to the group of 2×2 matrices with unit determinant $SL(2, \mathbb{R})$.

Unlike the above given examples, the actions of certain one-parameter groups cannot be obtained from linear transformations. These are non-linear one-parameter Lie groups and they do not have faithful matrix subgroup representations. For this category of Lie groups, we must define a broader characterization of their one-parameter Lie groups obtained by way of the exponential map. Let G be a Lie group with Lie algebra $T_{Id}G$, then the exponential map is defined by

$$
\begin{aligned}
exp_{Id}: \ T_{Id}G & \rightarrow & G \\
V & \mapsto & \gamma_V(1) \ ,
\end{aligned}
$$

where $\gamma_V : [0, +\infty) \rightarrow G$ is the constant-speed geodesic emanating from the identity element Id with $\dot{\gamma}_V(0) = V$. Hence this geodesic is precisely a one-parameter subgroup of G characterized by $\gamma_V(t) = exp_{Id}(tV)$. An example of a non-linear one-parameter Lie group acting on $\mathbb{R}^2$ is given by

$$
X = \frac{x}{1 + \lambda x} \ ; \ Y = (1 + \lambda x)^2 y \ .
$$

For a general one-parameter group, the quantities $\frac{dX}{d\lambda}|_{\lambda=0} := \xi(x, y)$ and $\frac{dY}{d\lambda}|_{\lambda=0} := \eta(x, y)$ are called the infinitesimal generators of the group, while

$$
X = x + \lambda.\xi(x, y) + O(\lambda^2) \ ; \ Y = y + \lambda.\eta(x, y) + O(\lambda^2)
$$

is referred to as the infinitesimal form of the group. More formally, an infinitesimal generator is a vector in the Lie algebra often represented as

$$
\xi \frac{\partial}{\partial x} + \eta \frac{\partial}{\partial y}
$$

for ξ, η as defined above. Given the infinitesimal form, we can deduce the global form by integrating the autonomous system of differential equations

$$
\frac{dX}{d\lambda} = \xi(X, Y) \ ; \ \frac{dY}{d\lambda} = \eta(X, Y)
$$

33

subject to the initial conditions $X|_{\lambda=0} = x$ and $Y|_{\lambda=0} = y$. By making the one-form $d\lambda$ subject in both equations of the system and equating the other terms, this system also helps to obtain the representation of the associated *group orbits* in $\mathbb{R}^2$.

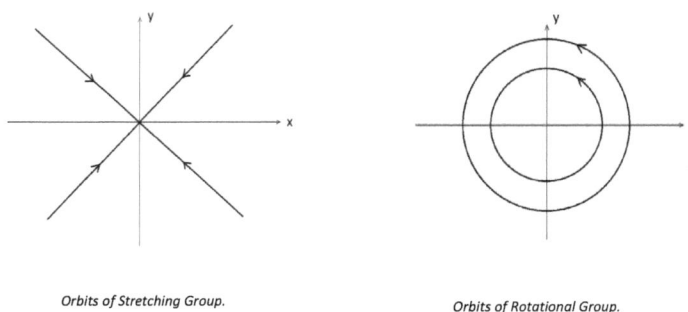

Orbits of Stretching Group.

Orbits of Rotational Group.

The orbit of a point $\mathbf{x}$ under group $\{P_\lambda\}$ denoted

$$\{P_\lambda\}\mathbf{x} := \{P_\lambda(\mathbf{x}) \mid \lambda \in \mathbb{R}\} \,,$$

is the collection of points to which $\mathbf{x}$ can be moved by elements of $\{P_\lambda\}$.

More succinctly, the group orbits may be obtained by integrating the differentials

$$\frac{dx}{\xi(x,y)} = \frac{dy}{\eta(x,y)} \,.$$

Given any one-parameter group, there exist functions $u(x,y)$ and $v(x,y)$ such that the global form of the group becomes

$$u(X,Y) = u(x,y) \;;\; v(X,Y) = v(x,y) + \lambda.$$

The function $u(x,y)$ is called a group invariant while (u,v) are referred to as the canonical coordinates of the group. If $u(x,y)$ is an invariant then so also is any function $\phi(u)$ and u may be found by eliminating the parameter

λ. Moreover, the group orbits of any one-parameter group each have the equation $u(x, y) = c$ where c is a real constant. To find v, for some constant x_0, define

$$\psi(x, u) := \int_{x_0}^{x} \frac{dt}{\xi(t, y(t, u))}$$

to obtain $v(x, y) = \psi(x, u(x, y))$.

3.2 Obtaining Invariant Solutions to D.E.'s by Symmetry Groups

A differential equation is said to accommodate a one-parameter Lie group if the action of that group on the domain of definition of the differential equation leaves the equation invariant. We will immediately demonstrate this concept with a nonlinear and inseparable first order O.D.E. Consider the D.E.

$$\frac{dy}{dx} = \frac{x^2 + y^2}{xy} \cdots (1)$$

defined on the open subset of $\mathbb{R}^2$ where x and y are both different from zero. By inspection, we recognize that this equation accommodates the stretching group $X = e^{\lambda}x$, $Y = e^{\lambda}y$ because the D.E.

$$\frac{dY}{dX} = \frac{X^2 + Y^2}{XY}$$

is equivalent to O.D.E (1); which is to say that the action of the stretching group leaves (1) invariant. We now introduce a group invariant which is chosen to eliminate λ : we set

$$u(X, Y) = u(x, y) = \frac{y}{x} .$$

Consequently,

$$du = \frac{\partial u}{\partial y}dy + \frac{\partial u}{\partial x}dx = \frac{dy}{x} - \frac{ydx}{x^2}$$

$$\Rightarrow dy = \frac{x^2 du + ydx}{x} = xdu + udx .$$

Substituting in (1), we have

$$\frac{xdu + udx}{dx} = \frac{x^2 + x^2u^2}{x^2u} = \frac{1 + u^2}{u}$$

$$\Rightarrow ux\frac{du}{dx} + u^2 = 1 + u^2$$

$$\Rightarrow \frac{du}{dx} = \frac{1}{ux}$$

$$\Rightarrow udu = \frac{dx}{x} \; .$$

Hence O.D.E (1) has become separable. We remark that in the case of any already separable first order O.D.E, employing an admitted symmetry transformation group yields a differential equation of greater or comparable difficulty in solving to the original.

The method of group invariant solutions can be extended to P.D.E's as illustrated by the following example. Consider the diffusion equation for $c(x,t)$,

$$\frac{\partial c}{\partial t} = \frac{\partial^2 c}{\partial x^2} \cdots (2)$$

subject to the boundary conditions

$$t \geq 0 \; , \; x \in \mathbb{R} \; , \; c(x,0) = c_0\delta(x).$$

This P.D.E is invariant under the transformation

$$P_\lambda : (x,t,c) \mapsto (X,T,C) \; ; \; X = e^\lambda x \; , \; T = e^{2\lambda}t \; , \; C = e^{-\lambda}c.$$

Observe that we have used an elementary property of the Dirac delta function δ, which is $\delta(\alpha x) = \alpha^{-1}\delta(x)$ for positive α, so that our solutions $c(x,t)$ vanish as $x,t \to \infty$. We can confirm P_λ to be the action of a one-parameter Lie group on Euclidean 3-space. The conditions to be satisfied for this case are similar to that for the plane, with the main difference being the number of arguments in the domain.

Given that $c = \phi(x,t)$ is a solution to (2), then we also have $C = \phi(X,T)$. This is true if ϕ has the functional form $\phi(x,t) = t^{-\frac{1}{2}}\psi(xt^{-\frac{1}{2}})$. Setting the group invariant $\zeta = xt^{-\frac{1}{2}}$, we are then able to transform (2) into the following ordinary differential equation;

$$2\psi''(\zeta) + \zeta\psi'(\zeta) + \psi(\zeta) = 0$$

which can be integrated once to obtain

$$2\psi'(\zeta) + \zeta\psi(\zeta) = A$$

for a constant A of integration. We may now integrate this linear first order O.D.E to obtain

$$\psi(\zeta) = exp\left(\frac{-\zeta^2}{4}\right)\int^{\zeta}\frac{A}{2}exp\left(\frac{\tau^2}{4}\right)d\tau + B.exp\left(\frac{-\zeta^2}{4}\right)$$

for a further constant of integration B. This result may only vanish as $t \to \infty$ if $A = 0$ so that

$$\psi(\zeta) = B.exp\left(\frac{-\zeta^2}{4}\right).$$

Hence, we have solved (2) completely by the one-parameter group invariant method upon retrieving the original variables and the boundary condition $c(x,0) = c_0\delta(x)$. Admitted symmetry groups of a D.E. which are not immediately realizable upon inspection can always be determined using the more general algebraic technique of vector field prolongations [9] (p.101 - 130). Also, more recent developments in P.D.E theory involve a wider view of conservation laws in solving diverse equations. The above two statements elude to grounds for further development of Sophus Lie's theory, but our next point of focus is on the relevance of Lie groups in simplifying second order linear O.D.E's (*LODEs*).

3.2.1   Symmetry Groups of Linear Second Order O.D.E's

It is important to consider transformations which preserve the order and linearity of this class of equations for the purpose of effectively implementing the tool of one-parameter symmetry groups. We will briefly discuss the transformation into normal form before discussing the most general transform for this purpose, namely, the Kummer-Liouville transform. Since the superposition principle is always applicable to linear differential equations, we will concentrate on the homogeneous form

$$y'' + a_1(x)y' + a_0(x)y = 0 \cdots (3)$$

We are guaranteed of existence of solutions to (3) whenever

$$a_1(x) \in C^1(I), a_0(x) \in C(I)$$

for an open, non-empty subinterval I of the real number line.

We may reduce (3) into normal form by changing the dependent variable to $y*$ where

$$y = (y*)exp\left[\frac{-1}{2}\int^x a_1(t)dt\right]$$

and the result of this transform is

$$\frac{d^2y*}{dx^2} + \left(a_0(x) - \frac{a_1'(x)}{2} - \frac{(a_1(x))^2}{4}\right)y* = 0 \cdots (3')$$

We will now examine which linear one-parameter symmetries of $(3')$ can be realized upon inspection. First, we set the coefficient of $y*$ to be

$$p(x) = \left(a_0(x) - \frac{a_1'(x)}{2} - \frac{(a_1(x))^2}{4}\right)$$

for convenience. Assuming $(3')$ accommodates a linear symmetry group;

$$G: \quad D \subseteq \mathbb{R}^2 \quad \to \quad \mathbb{R}^2$$
$$(x, y*) \quad \mapsto \quad \begin{pmatrix} G_{11} & G_{12} \\ G_{21} & G_{22} \end{pmatrix}\begin{pmatrix} x \\ y* \end{pmatrix} = \begin{pmatrix} X \\ Y \end{pmatrix}$$

then

$$X = G_{11}x + G_{12}y* \ , \ Y = G_{21}x + G_{22}y * \ .$$

Hence, $\dfrac{dY}{dX} = \dfrac{G_{21} + G_{22}(\frac{dy*}{dx})}{G_{11} + G_{12}(\frac{dy*}{dx})}$

and $\dfrac{d^2Y}{dX^2} = \dfrac{d}{dx}\left(\dfrac{dY}{dX}\right)\left(\dfrac{dx}{dX}\right) = (G_{22}G_{11} - G_{12}G_{21})\left(\dfrac{d^2y*}{dx^2}\right)\left(\dfrac{dx}{dX}\right)^3 \ .$

We desire that

$$\frac{d^2Y}{dX^2} + p(X)Y = 0$$

$$\Rightarrow p(X)(G_{21}x + G_{22}y*)\left(G_{11}^3 + 3G_{11}^2G_{12}\left(\frac{dy*}{dx}\right) + 3G_{11}G_{12}^2\left(\frac{dy*}{dx}\right)^2 + G_{12}^3\left(\frac{dy*}{dx}\right)^3\right)$$

$$+ (DetG)\frac{d^2y*}{dx^2} = 0 \ .$$

The following requirements for the invariance of $LODE(3')$ under G become evident:

$i.) G_{12} = 0$

$ii.) \dfrac{d^2y*}{dx^2} + p(X) \left[\dfrac{G_{11}^3 G_{21}x + G_{11}^3 G_{22}y*}{DetG} \right] = 0$

which together imply that

$$\frac{d^2y*}{dx^2} + \frac{G_{11}^2 G_{21}}{G_{22}}[p(G_{11}x)]x + G_{11}^2[p(G_{11}x)]y* = 0 .$$

Comparing this to (3'), we establish the final requirements:

$$G_{21} = 0 ; \ G_{11}^2[p(G_{11}x)] = p(x).$$

Thus, there are two cases to consider; the first being that $G_{11} \equiv 1$ (**I**) and the other being that G_{11} depends on a parameter (**II**).

(**I**)In the first case $G_{11} \equiv 1$, we have a scaling one-parameter group, whose matrix representation is

$$G = \begin{pmatrix} 1 & 0 \\ 0 & e^\lambda \end{pmatrix}$$

so that $X = x$, $Y = e^\lambda y*$ and a canonical coordinate for this transformation is $v(x, y*) = lny*$. Using this coordinate transformation in (3'), we have

$$y* = e^v ; \ (y*)' = v'e^v ; (y*)'' = (v'' + (v')^2)e^v$$

and then
$$v'' + (v')^2 + p(x) = 0.$$

Setting $z = v'$, we arrive at a non-linear, non-homogenous and usually inseparable Riccati differential equation of the first order:

$$z' = -z^2 - p(x).$$

Despite having reduced the order of $LODE(3)$ by one, our current scope ultimately transcends resolving this Riccati equation.

(**II**) In the second case where G_{11} is not fixed, we must have $p(x) = \dfrac{k}{x^2}$ where k is a real constant. This gives rise to a double parameter Lie group admittance upon inspection in normalized second order LODE's;

$$\frac{d^2y*}{dx^2} + \frac{k}{x^2}y* = 0$$

when implementing linear symmetry methods. Apart from the scaling group of the dependent variable identified in (**I**), this equation accommodates scalings of the independent variable : $x \mapsto e^{\mu}x$, which unfortunately, do nothing to simplify the state of ($3'$).

Despite these findings, the (linear) group invariance method is not maximally useful for solving (3) after transforming to normal form at this stage. However, we should acknowledge the interest generated by equations of the normal form ($3'$) when the function p is periodic in x, as highlighted by Berkovich and Rozov [7]. We remark also that if $\varphi(x)$ is a solution to ($3'$) then a second linearly independent solution is given by

$$\varphi(x) \int^x \frac{dt}{(\varphi(t))^2} \quad .$$

3.2.2 Kummer-Liouville Transformations of second order linear O.D.E's

By the *Stäckel − Lie* theorem [7], the Kummer-Liouville (KL) transformation is the most general point transformation that preserves the order and linearity of our *LODE*

$$y'' + a_1(x)y' + a_0(x)y = 0 \cdots (3).$$

The KL transform is given by

$$y = v(x)z \; , \; dt = u(x)dx \cdots (KL); \quad u, v \in C^2(I), \; uv \neq 0 \; \forall x \in I$$

which rearranges (3) to be of the form

$$\ddot{z} + b_1(t)\dot{z} + b_0(t)z = 0 \cdots (4); \quad b_1(t) \in C^1(J) \; , \; b_0(t) \in C(J)$$

where J is an open, non-empty sub-interval of the real number line. For clarity, we will use the prime sign ($'$) to denote differentiation with respect to x and an overset dot to denote differentiation with respect to t. Observe that we need the following three to occur in order to obtain (4) from (3) by way of transform (KL).

(i.) According to *Mammana's Theorem* [7], we must have the non-commutative factorization

$$Ly = \left(D - \frac{v'}{v} - \frac{u'}{u} - r_2(t)u\right)\left(D - \frac{v'}{v} - r_1(t)u\right)y = 0; \ D = \frac{d}{dx}$$

where $r_1(t)$ and $r_2(t)$ satisfy the Riccati equations:

$$\dot{r_1} + r_1^2 + b_1(t)r_1 + b_0(t) = 0 \ ; \ \dot{r_2} - r_2^2 - b_1(t)r_2 + \dot{b_1} - b_0(t) = 0$$

(ii.) $-2v'v^{-1} - u'u^{-1} + b_1(t)u = a_1(x)$

(iii.) $v'' + a_1v' + a_0v - b_0(t)u^2v = 0$.

The reduction of (3) to (4) was posed as Kummer's problem, which was to find the set of all KL transformations that could do this. It is known that Kummer's problem is always solvable. As a combination of the above three requirements for the KL transform, we get that (3) can be reduced to (4) if and only if the following two conditions are satisfied

$$v(x) = |u(x)|^{-\frac{1}{2}}exp\left(-\frac{1}{2}\int a_1(x)dx\right)exp\left(\frac{1}{2}\int b_1(t)dt\right) \cdots (E)$$

$$\frac{1}{2}\frac{t'''}{t'} - \frac{3}{4}\left(\frac{t''}{t'}\right)^2 + B_0(t)t'^2 = A_0(x) \cdots (E_2)$$

where

$$A_0(x) = a_0 - \frac{1}{4}a_1^2 - \frac{1}{2}a_1' \ ; \ B_0(t) = b_0 - \frac{1}{4}b_1^2 - \frac{1}{2}\dot{b_1}$$

are respectively called the semi-invariants of (3) and (4). We solve (ii.) over v in order to get (E) and then we substitute (iii.) by (E) using the relation $u = t'$ to get (E_2). At this juncture, it becomes clear that the transform from (3) into normal form $(3')$ is only a particular case of the (KL) transform with the functions

$$u(x) \equiv 1 \ ; \ v(x) = exp\left(-\frac{1}{2}\int a_1(x)dx\right) \ ; \ z = y * \ .$$

At the crux, we wish to reduce (3) to a LODE of autonomous form, that is, one with constant coefficients;

$$\ddot{z} + b_1\dot{z} + b_0z = 0 \cdots (4').$$

This can be factorized either through the noncommutative operators of the first order -

$$Ly \equiv \left(D - \frac{v'}{v} - \frac{u'}{u} - r_2 u \right) \left(D - \frac{v'}{v} - r_1 u \right) y = 0 \; ;$$

or through the commutative operators of first order -

$$\frac{1}{u^2} Ly \equiv \left(\frac{1}{u} D - \frac{v'}{uv} - r_2 \right) \left(\frac{1}{u} D - \frac{v'}{uv} - r_1 \right) y = 0$$

where r_1, r_2 are roots of the characteristic equation; $r^2 + b_1 r + b_0 = 0$.

We remark that (3) can be reduced to (4') by transform KL if and only if the following occur.

(i.) (3) admits a certain one-parameter Lie group

(ii.) $u(x)$ satisfies $\dfrac{1}{2} \dfrac{u''}{u} - \dfrac{3}{4} \left(\dfrac{u'}{u} \right)^2 + \left(\dfrac{-1}{4} b_1^2 + b_0 \right) u^2 = A_0(x)$

(iii.) $u''' - 6 \dfrac{u'u''}{u} + 6 \dfrac{(u')^3}{u^2} + 4A_0 u' - 2A_0' u = 0$

(iv.) The multiplier v and the kernel u of the KL transform are related through the formulas

$$v(x) = |u(x)|^{-\frac{1}{2}} exp \left(\frac{-1}{2} \int a_1(x) dx \right) exp \left(\frac{1}{2} b_1 \int u dx \right) \; ;$$

$$v'' + a_1 v' + a_0 v - b_0 u^2 v = 0$$

(v.) The resolvent of (3) is given by the function

$$R(x) = exp \left(- \int a_1 dx \right) |u|^{-1}$$

and it satisfies

(vi.) $R''' + 3a_1 R'' + (4a_0 + a_1' + 2a_1^2) R' + (2a_0' + 4a_0 a_1) R = 0$.

The one-parameter group existence follows from the reducibility of (3) to autonomous form (4'), as will soon be discussed. Condition (iii.) can be obtained from (ii.) by calculation, and then (vi.) can be obtained from (iii.) by way of the resolvent function

$$R(x) := y_1(x) y_2(x).$$

To be substituted in the resolvent function, we have linearly independent solutions of (3) given by

$$y_{1,2}(x) = |u|^{-\frac{1}{2}} exp\left(-\frac{1}{2}\int a_1 dx\right) exp\left(\pm\sqrt{\frac{b_1^2}{4} - b_0}\int u dx\right)$$

for the case of KL transform to autonomous form (4′).

3.2.3 Admittance of a Special Lie Group by Second-Order LODE's

The requirements $(i) - (vi)$ for the KL transform from (3) to (4′) suggest an iterative approach to find the kernel $u(x)$, such as the algorithm $REDUCE$ proffered by L. Berkovich and F. Berkovich [8]. Given the existence of $u(x)$, it can be effectively proven by the method of prolongation of the infinitesimal generators that (3) admits a linear symmetry group of dimension eight; corresponding specifically to the special Lie group

$$SL(3, \mathbb{R}) = \{A \in GL(3, \mathbb{R}) \mid DetA = 1\}.$$

Examining a few properties of this set as a manifold, we observe that $\mathcal{M}_{3\times3}(\mathbb{R})$ is isometrically isomorphic to $\mathbb{R}^9$, so that the set $SL(3, \mathbb{R})$ may rightly be considered to have the structure of a non-simply connected hypersurface in the ambient space $\mathcal{M}_{3\times3}(\mathbb{R})$. Recalling the KL substitutions $y = vz$, $t = \int u dx$ in (4′), we observe that

$$y''+y'\overbrace{\left(\frac{-2v'}{v} - \frac{u'}{u} + b_1 u\right)}^{a_1(x)} +y\overbrace{\left(\frac{2(v')^2}{v^2} - \frac{v''}{v} + \frac{v'u'}{vu} - \frac{b_1 uv'}{v} + b_0 u^2\right)}^{a_0(x)} = 0\cdots(5).$$

First and foremost, we reckon that (5) admits the infinitesimal generator

$$\chi_1 = \frac{1}{u}\frac{\partial}{\partial x} + \frac{v'}{uv}y\frac{\partial}{\partial y}.$$

Consequently, in seeking an invariant for (5), we integrate the differentials

$$u dx = \frac{uv}{yv'}dy \ ,$$

identifying that

$$v(x) = |u(x)|^{-\frac{1}{2}} exp\left(\frac{-1}{2}\int a_1(x)dx\right) exp\left(\frac{1}{2}b_1\int udx\right).$$

Integrating the differentials gives us $\frac{y}{v} = constant$, so $\frac{y}{v}$ is an invariant. Solving for the second canonical coordinate of χ_1, we get it to be

$$\int_{x_0}^{x}\frac{d\tau}{\frac{1}{u}} = \int udx,$$

which is expected, since the transformation from (5) with the substitutions $\frac{y}{v} := z$ and $\int udx := t$ gives the autonomous O.D.E (4′).

The eight dimensions of $SL(3,\mathbb{R})$ permit seven other independent infinitesimal generators admitted by (5) besides χ_1 which are mentioned as follows;

$$\chi_2 = v\frac{\partial}{\partial y} \qquad\qquad \chi_3 = \chi_1\int udx$$

$$\chi_4 = \chi_2\int udx \qquad\qquad \chi_5 = \frac{y}{v}\chi_1$$

$$\chi_6 = \frac{y}{v}\chi_2 \qquad\qquad \chi_7 = \left(\int udx\right)^2\chi_1 + \left(\frac{y}{v}\int udx\right)\chi_2$$

$$\chi_8 = \left(\frac{y}{v}\int udx\right)\chi_1 + \left(\frac{y}{v}\right)^2\chi_2.$$

Following up on the prior observed immediate admittance of the scaling group by (3), it can easily be verified that the infinitesimal generator χ_6 corresponds to this one-parameter group.

As an alternative to arrive at the general solution, we may reduce (5) after expressing it in normal form using the point transformation

$$y = (y*).exp\left[-\frac{1}{2}\int a_1dx\right].$$

In so doing, we observe that

$$(y*)'' + \left(b_0u^2 + \frac{u''}{2u} - \frac{3(u')^2}{4u^2} - \frac{b_1^2u^2}{4}\right)y* = 0 \cdots (5')$$

where the coefficient of $y*$; $A_0 = b_0u^2 + \frac{u''}{2u} - \frac{3(u')^2}{4u^2} - \frac{b_1^2u^2}{4}$ has been previously identified as a semi-invariant of (3). The transformation to normal form does not alter the invariants of the differential equation, so it still accommodates the invariant

$$\frac{y}{v} = \frac{y*}{|u|^{-\frac{1}{2}}.exp(-\frac{b_1}{2}\int udx)} := z$$

realized from the infinitesimal generator χ_1. The advantage of using point transformations, such as the KL transform, prior to finding point symmetries is a possible re-structuring of the infinitesimal generators; which can enable study of a given D.E. in greater depth.

In general, computing all the invariants for a multiple-parameter group of transformations can get quite complicated. Each invariant of (5) must be a *joint invariant* of $\chi_1 - \chi_8$, meaning that any invariant of the system can be expressed as a function of the invariants of all eight generators. Because we need only the invariant $\frac{y}{v} := z$ and the other canonical coordinate $\int udx := t$ obtained from χ_1 to complete the reduction to autonomous form (4′), this infinitesimal generator is recognized foremost.

After making the substitutions

$$\frac{y*}{|u|^{-\frac{1}{2}}.exp(-\frac{b_1}{2}\int udx)} := z \text{ and } \int udx := t,$$

equation (5′) is readily reduced to the autonomous form

$$\ddot{z} - b_1\dot{z} + b_0 z = 0.$$

The solution, substituting back for $y*$, yields

$$y*_{1,2}(x) = |u|^{-\frac{1}{2}}exp\left(\pm\sqrt{\frac{b_1^2}{4} - b_0}\int udx\right)$$

and finally the linearly independent solutions of (3)

$$y_{1,2}(x) = |u|^{-\frac{1}{2}}exp\left(-\frac{1}{2}\int a_1 dx\right)exp\left(\pm\sqrt{\frac{b_1^2}{4} - b_0}\int udx\right).$$

This implies the reduction of the initial problem (3) to the simpler problem of integrating a_1 and the kernel of the KL transform to finalize the solution.

As illustrated, the introduction of the the kernel (u) and the multiplier (v) of the KL transform are instrumental in realizing the $SL(3, \mathbb{R})$ symmetry and general solution for LODE (3). After coming fully to terms with the reducibility concept, attention then shifts to the efficiency of algorithms for computing the kernel and multiplier of the KL transform. This may be done successfully with aid of computer algebra, which is also a handy alternative for obtaining the point symmetries of higher order differential equations. Even though not all differential equations can be simplified by symmetry considerations (since not all differential equations admit symmetry groups), the method of invariant Lie group transformations is perhaps one of the most convenient techniques to be incorporated when available.

CHAPTER **4**

P.D.E FORMULATIONS FROM VARIATIONAL
PROBLEMS

Preview. The calculus of variations applied in multivariate problems can give rise to several classical Partial Differential Equations (P.D.E's) of interest. In this chapter, we aim to formulate such equations arising from the viewpoint of optimization of energy functionals on smooth Riemannian manifolds. Target domains are taken as appropriate subsets of Sobolev spaces, with briefings on analytical implications and approaches proffered.

4.1   Introduction.

In due course, we will invoke two fundamental tools during formulation which are outrightly stated in this section. The first is a general theorem in functional optimization theory and the second is a lemma of variational calculus.

Optimization Theorem - [2] (p. 155) Let E be a real reflexive Banach space, and the functional $f : E \to \mathbb{R} \cup \{+\infty\}$ be convex, lower-semi continuous and proper. Then,
i.) for any non-empty $K \subset E$ that is weakly compact (closed, convex and norm-bounded) $\exists\, \overline{x} \in K$ such that $f(\overline{x}) = \min_{x \in K} f(x)$;

ii.) if in addition f is coercive, $\exists\ \overline{x} \in E$ such that $f(\overline{x}) = \min\limits_{x\in E} f(x)$.

Lemma of The Calculus of Variations - Let $\Omega \subset \mathbb{R}^n$ be a regular, orientable and bounded submanifold-with-boundary. Let $h \in C(\Omega)$ and assume that

$$\int_\Omega h.\phi d\mu = 0$$

for all $\phi \in C_0^1(\Omega)$. Then $h \equiv 0$ on Ω, where Ω is of dimension m and μ is the geometric m-volume on Ω.

We now proceed to prove this lemma by contradiction. Assume $|h(x_0)| > 0$ for some $x_0 \in \Omega \backslash \partial\Omega$ ($\partial\Omega$ is the boundary of $\overline{\Omega}$ defined in the geometric sense). Then for some $\delta > 0$ we have $d(x_0, \partial\Omega) < \delta$ and $|h(x)| \geq \dfrac{|h(x_0)|}{2}$ on $B(x_0, \delta) \cap \Omega$.
Setting

$$\phi(x) = \begin{cases} (\delta^2 - ||x - x_0||^2)^2; & \text{if } x \in B(x_0, \delta) \cap \Omega \\[2mm] 0; & \text{if } x \in \Omega \backslash B(x_0, \delta) \end{cases}$$

then $\phi \in C_0^1(\Omega)$. To confirm this claim, we justify differentiability of ϕ by first taking Ω to be $\mathbb{R}^n$. Differentiability of this function is obvious everywhere but on $\partial B(x_0, \delta)$. The gradient of this function for $x \in B(x_0, \delta)$ is given by $\phi'(x) = 4(||x - x_0||^2 - \delta^2).(x - x_0)$. Thus, for any sequence $\{x_j\}_{j\in\mathbb{N}} \subset \mathbb{R}^n \backslash \partial B(x_0, \delta)$ approaching some point $x \in \partial B(x_0, \delta)$, we get $\phi'(x_j)$ to be approaching zero. This gives us that the gradient function ϕ' is continuous on $\mathbb{R}^n$. Moreover,

$$|\int_\Omega h(x)\phi(x)d\mu| \geq \frac{|h(x_0)|}{2} \int_{B(x_0,\delta)\cap\Omega} \phi(x)d\mu > 0$$

which contradicts our initial assumption.

   If Ω is not open, then it becomes expedient for us to define the gradient function intrinsically, using a covariant derivative for ϕ on Ω. In this event, we take

$$\phi'|_{T\Omega} = \nabla\phi = (\langle\phi', E_1\rangle, \langle\phi', E_2\rangle, \cdots, \langle\phi', E_m\rangle) := (\nabla_{E_i}\phi)_{i=1}^m$$

where $\{E_1, E_2, \cdots, E_m\}$ is an orthonormal basis of functions for the tangent spaces to Ω at each point and $[E_1, E_2, \cdots, E_m]_p$ is the usual orientation for

48

$T_p\Omega$ at each $p \in \Omega$. We can draw the same conclusions by adjusting certain conditions of the lemma, such as taking Ω to be a submanifold-without-boundary and using $\phi \in C^1(\overline{\Omega})$ as the test functions, when appropriate.

The function spaces $C^1(\overline{\Omega})$ and $C_0^1(\Omega)$ are not reflexive, which will prompt us to consider instead weak formulations of the optimization problems during computations. That is to say, we will target solutions in larger reflexive Sobolev spaces, reckoning with the fact that $C^1(\overline{\Omega})$ is dense in $W^{k,p}(\Omega)$ for any natural k and for $1 < p < \infty$. Occasionally, weak solutions also turn out to be solutions of the classical problems. Following our presentation of the fundamental tools, we now give two methods of formulation of classical PDE's associated to optimization of differentiable functionals.

4.2 Methods of Formulation

4.2.1 Method 1

Let $\Omega \subseteq \mathbb{R}^n$ be a connected, orientable and bounded submanifold of class C^2, and we are to minimize the functional

$$
\begin{aligned}
E : \quad V &\longrightarrow \mathbb{R} \\
v &\longmapsto \int_\Omega F(x, v, \nabla v) d\mu
\end{aligned}
$$

on a subset $V \subseteq C^1(\overline{\Omega})$ for a sufficiently regular function F. Let $\overline{v}$ be a minimizer of E on V. Then for some real positive number r, we have

$$E(\overline{v}) \leq E(v) \ \forall v \in V \cap B(\overline{v}, r) \ .$$

We will choose V such that it accommodates appropriate tangent cones, which are variations of the form $v + \tau\phi$ for $v \in V$ and $\phi \in C_0^1(\Omega)$ (resp. $C^1(\overline{\Omega})$). There exists a positive real number δ such that for any $\tau \in (-\delta, \delta)$ we have

$$\overline{v} + \tau\phi \in B(\overline{v}, r) \ .$$

Defining $\gamma(\tau) := E(\overline{v} + \tau\phi)$, then 0 is a minimizer of γ in $(-\delta, \delta)$ which means

$$\gamma'(0) = 0 \ \Rightarrow \ E'(\overline{v}).\phi = 0 \ .$$

We find the derivative of E at $\overline{v}$ in the direction of ϕ;

$$
\begin{aligned}
E'(\overline{v}).\phi &= \lim_{\alpha \to 0} \left(\frac{E(\overline{v} + \alpha\phi) - E(\overline{v})}{\alpha} \right) \\
&= \int_{\Omega} \lim_{\alpha \to 0} \frac{F(x, \overline{v}(x) + \alpha\phi(x), \nabla\overline{v}(x) + \alpha\nabla\phi(x)) - F(x, \overline{v}(x), \nabla\overline{v}(x))}{\alpha} d\mu \\
&= \int_{\Omega} F'(x, \overline{v}, \nabla\overline{v}); (0, \phi, \nabla\phi) \, d\mu \\
&= \int_{\Omega} [F_v(x, \overline{v}, \nabla\overline{v}).\phi + F_{\nabla v}(x, \overline{v}, \nabla\overline{v}).\nabla\phi] \, d\mu \ .
\end{aligned}
$$

The regularity of F at $\overline{v}$ is necessary for passing the limit into the above integral as done above. This gives us uniform convergence of the integrand corroborated by the mean value inequality, since the integrand equals $F'(\zeta); (0, \phi, \nabla\phi)$ for some $\zeta \in \mathbb{R}^m \times \mathbb{R} \times \mathbb{R}^m$, with $||F'||$ having a finite upper bound independent of α in a neighborhood of ζ.

We hereby explain certain used notations by again having $\{E_i\}_{i=1}^m$ as an orthonormal basis of functions for the tangent bundle $T\Omega$ at each point of the manifold. We represent the gradient functions intrinsically to Ω :

$$
\nabla v = (\nabla_{E_i} v)_{i=1}^m \text{ and } \nabla \phi = (\nabla_{E_i} \phi)_{i=1}^m \ .
$$

This way, $F_{\nabla v}$ has m scalar components, each of which we will denote $F_{v_{E_i}}$. We then implement Green's theorem of multivariate integration to re-evaluate a term in the formulation

$$
\int_{\Omega} F_{\nabla v}.\nabla\phi d\mu
$$
$$
= \int_{\Omega} (F_{v_{E_i}})_{i=1}^m.(\nabla_{E_i}\phi)_{i=1}^m d\mu
$$
$$
= -\int_{\Omega} (div F_{\nabla v}) \phi d\mu \ .
$$

As such, we assume in addition that F is of class C^2 and we get that $\forall \phi \in C_0^1(\Omega)$ (or $C^1(\overline{\Omega})$ if $\overline{\Omega}$ is without boundary) ,

$$
\int_{\Omega} (F_v(x, \overline{v}, \nabla\overline{v}) - div F_{\nabla v}) \phi \, d\mu = 0 \ \Rightarrow
$$

$$
F_v(x, \overline{v}, \nabla\overline{v}) = div F_{\nabla v}(x, \overline{v}, \nabla\overline{v}) \cdots (1)
$$

at any local or global minimizer $\overline{v}$ of E on V, which gives us a necessary optimality condition. If Ω is an open subset of $\mathbb{R}^n$, then the above is simply written as

$$F_v(x, \overline{v}, \nabla\overline{v}) = \sum_{i=1}^n \frac{\partial}{\partial x_i} F_{v_{xi}}(x, \overline{v}, \nabla\overline{v}) \ .$$

4.2.2 Method 2

Now, assume that $\Omega \subseteq \mathbb{R}^n$ is open, bounded with a regular topological boundary, and we are given the following functional to minimize;

$$E : \ V \subseteq C^1(\overline{\Omega}) \ \longrightarrow \ \mathbb{R}$$
$$v \ \longmapsto \ \int_\Omega F(x, v, \nabla v) dV - \int_{\partial\Omega} g(x, v) dS \ .$$

For this case, we work with test functions $\phi \in C^1(\overline{\Omega})$ because of the contributor to the functional from the boundary, and our formulation yields the necessary optimality condition

$$\int_\Omega \left[F_v(x, \overline{v}, \nabla\overline{v}).\phi + F_{\nabla v}(x, \overline{v}, \nabla\overline{v}).\nabla\phi \right] dV - \int_{\partial\Omega} g_v(x, \overline{v})\phi dS = 0$$

at any minimizer $\overline{v}$ of E in V. Of course, we need the functions F and g to be sufficiently regular (ideally, F to be of class C^2 and g to be of class C^1). Applying Green's theorem to a term in the above formulation;

$$\int_\Omega F_{\nabla v}.\nabla\phi \ dV = \int_{\partial\Omega} \left(\frac{\partial F_{\nabla v}}{\partial N}.\phi \right) dS - \int_\Omega \left(\sum_{i=1}^n \frac{\partial}{\partial x_i} F_{v_{xi}} \right).\phi \ dV$$

where N is the outward unit normal or Gauss map evaluated on $\partial\Omega$ and

$$\frac{\partial F_{\nabla v}}{\partial N} := \langle F_{\nabla v}, N \rangle \ .$$

Substituting this in our formulation, we get

$$\int_\Omega \left(F_v - \sum_{i=1}^n \frac{\partial}{\partial x_i} F_{v_{xi}} \right) \phi dV + \int_{\partial\Omega} \left(\langle F_{\nabla v}, N \rangle - g_v \right) \phi dS = 0$$

at any local or global mininizer $\overline{v}$ of E. By the fundamental lemma of variational calculus, we conclude -

$$\sum_{i=1}^n \frac{\partial}{\partial x_i} F_{\overline{v}_{x_i}} = F_{\overline{v}} \ \text{ in } \Omega;$$

$$\sum_{i=1}^{n} F_{\overline{v}_{x_i}} N_i = g_{\overline{v}} \quad \text{on } \partial\Omega \cdots (2)$$

which is a boundary value P.D.E problem of Neumann type. By slightly adjusting prior hypotheses, it is straightforward to generalize that the formulations from (1) and (2) give not only the necessary optimality conditions for minimizers, but for local critical points at large.

The given lemma of variational calculus applies to larger reflexive Sobolev spaces, and this we can infer from a generalization of the lemma called the du Bois - Reymond lemma. It gives us that for any $h \in W^{k,p}(\Omega)$ satisfying $\int_{\Omega} h.\phi d\mu = 0$ for all $\phi \in C_0^1(\Omega)$, then $h \equiv 0$ almost everywhere on Ω. However, it is most convenient to work with subsets of the reflexive Hilbert space $W^{k,2}(\Omega)$ because of continuity of the partial inner product -

$$\langle h, \phi \rangle = \int_{\Omega} h.\phi d\mu \quad \text{for } h, \phi \in W^{k,2}(\Omega)$$

and the availability of other analytical solution tools such as the Lax-Milgram theorem. Here, given $h \in W^{k,2}(\Omega)$ and $\int_{\Omega} h.\phi d\mu = 0$ for all $\phi \in W^{k,2}(\Omega)$, then $h \equiv 0$ almost everywhere on Ω. We will consider the weakly formulated methods in these larger reflexive Sobolev spaces in order to investigate existence and/or uniqueness of solutions to the optimization problems. Given the possibility of solution existence from the optimization theorem, we proceed to solve the weakly formulated P.D.E in a Sobolev space, and finally check whether the weak solutions are also classical solutions in $C^1(\overline{\Omega})$. It may turn out that the weakly formulated problem has a solution, while the classical problem does not. In the succeeding examples, we will illustrate the theoretical framework laid out above.

4.3 Examples

4.3.1 Example 1 : Perelman Entropy Functional

Let $S \subset \mathbb{R}^n$ be a regular, compact and connected hypersurface. A Perelman entropy functional on S is formally analogous to heat flow along the manifold and we give it by

$$E: \quad C^1(S) \longrightarrow \mathbb{R}$$
$$v \longmapsto \int_S (R + ||\nabla v||^2) exp(-v) \; dS$$

where R is the scalar curvature of S. In any setting, the functional E lacks coercivity, and usually it also lacks a global minimizer.

As an illustration of this statement, let S be a regular surface in $\mathbb{R}^3$ consisting only of elliptic points, in which case the scalar curvature equals twice the Gaussian curvature and the Perelman entropy is strictly positive for any v. We consider the problem in weak settings in order to investigate arguments using our optimization theorem. W can be any of the classical Sobolev spaces containing $C^1(S)$, and we see that $\inf_{v \in W} E(v) = 0$ by taking $||v||_W$ to infinity along the positive direction of constant functionals $v \equiv k$ where $k \in \mathbb{R}^+$ is a constant.

Nevertheless, the formulation of method 1 (subsection 2.1) above provides weak local critical points of E which we hereby discuss. This formulation gives us

$$-(R + ||\nabla \overline{v}||^2)exp(-\overline{v}) = 2div(exp(-\overline{v})\nabla \overline{v}) \quad \cdots (P_1)$$

at any critical point $\overline{v}$ of E. Observe that (P_1) is set as a non-linear second order P.D.E, as the divergence operation on the right hand side produces a second order differential of $\overline{v}$. As a simple computational illustration, we will have S to be the two-dimensional unit sphere $S^2 \subset \mathbb{R}^3$. This is a compact surface consisting only of elliptic points embedded in the real Euclidean 3-space with unit Gaussian curvature $K \equiv 1$ so that its scalar curvature is also constant : $R \equiv 2$. Solving for $\overline{v}$ in this case, we have

$$\left(2 + (\nabla_{E_1}\overline{v})^2 + (\nabla_{E_2}\overline{v})^2\right) exp(-\overline{v}) + 2div(exp(-\overline{v})\nabla \overline{v}) = 0 \implies$$

$$\left(2 + \sum_{j=1,2} \left(\frac{\nabla_{\Phi_{u_j}}\overline{v}}{||\Phi_{u_j}||}\right)^2\right) exp(-\overline{v}) + 2div\left(exp(-\overline{v})\left(\frac{\nabla_{\Phi_{u_j}}\overline{v}}{||\Phi_{u_j}||}\right)_{j=1,2}\right) = 0 \implies$$

$$\left(2 + \frac{1}{g_{jj}}\left(\frac{\partial f}{\partial u_j}\right)^2\right) exp(-f) + 2\sqrt{|g^{-1}|}\frac{\partial}{\partial u_j}\left(\sqrt{|g|}\exp(-f)\sqrt{\frac{1}{g_{jj}}}\frac{\partial f}{\partial u_j}\right) = 0$$

where $f = \overline{v} \circ \Phi$ and the parametrization

$$\Phi : U \subseteq \mathbb{R}^2 \implies S^2 \; ; \; (u_1, u_2) \mapsto \Phi(u_1, u_2) := p \in S^2$$

is used to pull back the differential equation to an open subset $U \subseteq \mathbb{R}^2$ and solve. Using the spherical co-ordinate system, we have

$$\begin{aligned}\Phi : \quad U = (0, 2\pi) \times (-\tfrac{\pi}{2}, \tfrac{\pi}{2}) \subset \mathbb{R}^2 &\longrightarrow S^2 \\ (u_1, u_2) &\longmapsto (\cos u_1 \cos u_2, \sin u_1 \cos u_2, \sin u_2)\end{aligned}$$

so that

$$g_{11} = \left\| \frac{\partial \Phi}{\partial u_1} \right\|^2 = cos^2 u_2 \ ; \ g_{22} = \left\| \frac{\partial \Phi}{\partial u_2} \right\|^2 = 1.$$

We have E_1, E_2 to be the unit vectors in the directions of Φ_{u_1} and Φ_{u_2} respectively. Observe that we have used the elementary property of directional derivatives; $\nabla_{\alpha X} Y = \alpha \nabla_X Y$ for a scalar field α. Hence the formulation in P_1 becomes

$$2 + \sec^2 u_2 \left(\frac{\partial f}{\partial u_1} \right)^2 - \left(\frac{\partial f}{\partial u_2} \right)^2 - 2 \sec u_2 \left(\frac{\partial f}{\partial u_1} \right)^2 + 2 \sec u_2 \frac{\partial^2 f}{\partial u_1^2} - 2 \tan u_2 \frac{\partial f}{\partial u_2} + 2 \frac{\partial^2 f}{\partial u_2^2} = 0.$$

We will set $f_{u_1} = 0$ for simplicity. In so doing, we obtain a second order non-linear O.D.E:

$$2 - [f'(u_2)]^2 - 2 \tan(u_2).f'(u_2) + 2f''(u_2) = 0.$$

Substitution of the variable $f'(u_2) = w$ gives an O.D.E of the first order. Although solutions exist on $(-\frac{\pi}{2}, \frac{\pi}{2})$, specified initial value solutions may not be bounded or unique on the interval because the tangent function is only continuous, and not uniformly continuous therein. For instance, we must use the initial condition $w(-\frac{\pi}{2}) = 0$ in order to avoid a singularity at the point $u_2 = -\frac{\pi}{2}$, but the solution obtained under this specification is unbounded because $\lim\limits_{u_2 \to \frac{\pi}{2}^-} w(u_2) = -\infty$. For this case, we can confirm that $\overline{v} \in L^1_{loc}(S^2)$, which is the largest of function spaces covered by the du Bois - Reymond lemma. As such, $\overline{v}$ is almost everywhere locally integrable since it is continuous except at $\Phi(u_1, \frac{\pi}{2})$, which is a point. However, continuous differentiability on the entire compact unit sphere is not obtainable.

The nature of a critical function $\overline{v}$ can be investigated using the second variation test of the functional E. We can judge the nature of $\overline{v}$ by considering variations of the form $\overline{v} + \epsilon\phi$ for $\epsilon > 0$ small enough to judge how E acts locally around $\overline{v}$. We may examine the most significant terms of an associated Taylor Series expansion for this purpose. The smoothness of the metric guarantees that E is (more than) twice differentiable, so that we can deduce the required terms as follows.

$$E(v + \epsilon\phi)$$
$$= \int_{S^2} F(x, v + \epsilon\phi, \nabla v + \epsilon\nabla\phi) dS$$
$$= \int_{S^2} F[(x, v, \nabla v) + \epsilon(0, \phi, \nabla\phi)] dS.$$

We now expand the integrand to get

$$F(x, v, \nabla v) + \epsilon F'(x, v, \nabla v); (0, \phi, \nabla \phi) + \frac{\epsilon^2}{2} F''[(x, v, \nabla v); (0, \phi, \nabla \phi)]; (0, \phi, \nabla \phi) + \mathcal{O}(\epsilon^3)$$

$$= F(x, v, \nabla v) + \epsilon F'(x, v, \nabla v); (0, \phi, \nabla \phi) + \frac{\epsilon^2}{2}(F_{vv}\phi^2 + 2\phi F_{v\nabla v}.\nabla \phi + [F_{\nabla v \nabla v}; \nabla \phi].\nabla \phi) + \mathcal{O}(\epsilon^3).$$

Therefore,

$$E(v + \epsilon \phi) - E(v) =$$
$$\int_{S^2} \left(\epsilon F'(x, v, \nabla v); (0, \phi, \nabla \phi) + \frac{\epsilon^2}{2}(F_{vv}\phi^2 + 2\phi F_{v\nabla v}.\nabla \phi + [F_{\nabla v \nabla v}; \nabla \phi].\nabla \phi) + \mathcal{O}(\epsilon^3) \right) dS$$

$$\Rightarrow E(\overline{v} + \epsilon \phi) - E(\overline{v}) =$$
$$\int_{S^2} \left(\frac{\epsilon^2}{2}(F_{vv}\phi^2 + 2\phi F_{v\nabla v}.\nabla \phi + [F_{\nabla v \nabla v}; \nabla \phi].\nabla \phi) + \mathcal{O}(\epsilon^3) \right) dS \mid_{\overline{v}}.$$

The sign of the result obtained above is determined by the leading term of the integrand; $\frac{\epsilon^2}{2}(F_{vv}\phi^2 + 2\phi F_{v\nabla v}.\nabla \phi + [F_{\nabla v \nabla v}; \nabla \phi].\nabla \phi)$. This is to say, if the critical function $\overline{v}$ is a strict local minimizer then $E(\overline{v}+\epsilon\phi) - E(\overline{v}) > 0$ for ϵ small enough, meaning $(F_{vv}\phi^2 + 2\phi F_{v\nabla v}.\nabla \phi + [F_{\nabla v \nabla v}; \nabla \phi].\nabla \phi)$ is positive for any non-zero test function $\phi \in C^1(S^2)$. By a similar line of reasoning, we can make the opposite conclusion in the event whereby $\overline{v}$ is a local maximizer of E. Critical points for which $E(\overline{v} + \epsilon\phi) - E(\overline{v})$ is neither positively nor negatively defined on any ϵ - neighborhood of $\overline{v}$ in the ambient Sobolev space are saddle points.

The procedure just outlined is that of examining the second variation of E. Indeed, we can readily view this as the infinite-dimensional analogue of the behavior of the second derivative of functionals, by way of the quadratic form

$$E''(v)(\phi)(\phi) = \int_{S^2} (F_{vv}\phi^2 + 2\phi F_{v\nabla v}.\nabla \phi + [F_{\nabla v \nabla v}; \nabla \phi].\nabla \phi)dS.$$

With the given analytical observation, we now make remarks based on the computation initiated above on the Perelman entropy functional.

$$E''(v)(\phi)(\phi)$$
$$= \int_{S^2} [(2 + ||\nabla v||^2)\phi^2 e^{-v} - 4e^{-v}\phi \langle \nabla \phi, \nabla v \rangle + 2||\nabla \phi||^2 e^{-v}]dS$$
$$= \int_{S^2} e^{-v}[(2 + ||\nabla v||^2)\phi^2 - 4\phi \langle \nabla \phi, \nabla v \rangle + 2||\nabla \phi||^2]dS \cdots (O_1)$$

Observe also that

$$2||\phi\nabla v - \nabla\phi||^2$$
$$= 2\langle\phi\nabla v - \nabla\phi, \phi\nabla v - \nabla\phi\rangle$$
$$= 2\phi^2||\nabla v||^2 - 4\phi\langle\nabla v, \nabla\phi\rangle + 2||\nabla\phi||^2 \geq 0 \ \cdots (O_2).$$

If everywhere on S, we have $||\nabla\overline{v}||^2 < 2$, then whenever $\phi \neq 0$ we also get

$$(2+||\nabla\overline{v}||^2)\phi^2 - 4\phi\langle\nabla\phi, \nabla\overline{v}\rangle + 2||\nabla\phi||^2 > 2\phi^2||\nabla\overline{v}||^2 - 4\phi\langle\nabla\overline{v}, \nabla\phi\rangle + 2||\nabla\phi||^2$$

$$\Rightarrow E''(\overline{v})(\phi)(\phi) > 0, \quad \text{from } (O_1) \text{ and } (O_2).$$

In such an event, the critical function $\overline{v}$ would be a strict local minimizer for the Perelman entropy. Using the weak solution obtained above, this situation can be created simply by truncating the parallels u_2^* for which $|w(u_2^*)| > \sqrt{2}$. The resulting manifold would be a connected spherical section with boundary, and the critical function would then be a classical solution to P_1.

If $||\nabla\overline{v}||^2 > 2$ on a dS-non-negligible subset S^* of S^2, then $\overline{v}$ would be a saddle point for E. This is confirmed by using an appropriate Urysohn test function ϕ, satisfying

$$\phi := exp(\overline{v}) \quad \text{on a non-negligible subset of } S^* \text{ and}$$
$$\phi = 0 \quad \text{whenever } ||\nabla\overline{v}||^2 < 2,$$

for which $E''(\overline{v})(\phi)(\phi) < 0$. It is easy to verify that the integrand of $E''(\overline{v})(\phi)(\phi)$ is negative whenever $||\nabla\overline{v}||^2 > 2$ and ϕ equals $exp(\overline{v})$.

E has no strict local maximizers as $E''(v)(\kappa)(\kappa) > 0$ for any non-zero constant function κ.

In the classical applications of Perelman's entropy, the functional v is time dependent and critical points of E with respect to time characterize steady *Ricci Solitons*. Therefore, Perelman's entropy can be regarded as a variational view point for the study of *Ricci flow*, which is an advanced current area of interest in pseudo-Riemannian geometry. It is also worthy of note that the same functional also occurs quite importantly in a branch of theoretical physics known as string theory.

4.3.2 Example 2: Dirichlet Energy Functional

Let $\Omega \subset \mathbb{R}^n$ be open, bounded and with C^1 topological boundary. The Dirichlet energy functional on Ω is given by

$$E : \quad V \subset C^1(\overline{\Omega}) \longrightarrow \mathbb{R}$$
$$v \longmapsto \int_\Omega ||\nabla v||^2 \, dV$$

where

$$V = \{v \in C^1(\overline{\Omega}) : v|_{\partial\Omega} = h\}$$

and h is a particular differentiable function defined on the compact set $\partial\Omega$. The classical problem is to minimize E over V, but we first consider the weak setting in the reflexive Sobolev space $W^{1,2}(\Omega) := H^1(\Omega)$ to effectively perform analysis using the given optimization theorem. In particular, this setting is made appropriate by continuous differentiability of the functional E on $H^1(\Omega)$.

Hence, the domain V' of E in this setting is the pre-image of the singleton $h \in L^2(\partial\Omega)$ under the continuous trace operator;

$$\gamma_0 : H^1(\Omega) \to L^2(\partial\Omega)$$

so V' is (norm-) closed in $H^1(\Omega)$. Moreover, the set V' is convex because $\lambda u + (1-\lambda)v \in V' \; \forall \lambda \in [0,1]$ and every $u, v \in V'$. The functional E is continuous, coercive and strictly convex on V'. For any function v_0 in V', it is easy to check that the set $B = \{v \in V' : E(v) \leq E(v_0)\}$ is bounded due to the coercivity of E, giving us existence of a minimizer for E on B and thus also on V'. The critical function $\overline{v}$ will exist uniquely due to strict convexity of E.

Given the minimizer $\overline{v} \in H^1(\Omega)$, the weak formulation for this problem is the following boundary value P.D.E:

$$\sum_{i=1}^n \frac{\partial^2 \overline{v}}{\partial x_i^2} = 0 \; \text{ in } \Omega;$$

$$\overline{v} = h \; \text{ on } \partial\Omega \quad \cdots (P_2).$$

This is obtained by implementing the formulation method 2 (subsection 2.2) with test functions in $H_0^1(\Omega)$ instead of $H^1(\Omega)$ because there is no contribution to the functional E from the boundary of Ω. Problem (P_2) is known as Laplace's equation, as

$$\sum_{i=1}^n \frac{\partial^2 \overline{v}}{\partial x_i^2} := \Delta\overline{v}$$

57

is called the Laplacian of $\overline{v}$. (P_2) is a linear second-order elliptic P.D.E of Dirichlet type, of which the solutions constitute an interesting category of functions. Its solutions are called harmonic functions and they are the chief ingredients in the study of potential theory. We will hereby give just a brief analysis of this formulation and possible solutions.

Symmetries of The Laplace Equation

One of the most efficient approaches to tackling Laplace's equation is exploiting its symmetries. This equation is known to accommodate the Lie groups of conformal transformations on $\mathbb{R}^n$, which are precisely the non-degenerate symmetry Lie groups of invariance transformations which preserve angles between the vectors in their domains. Any such group can be decomposed into one-parameter subgroups. Each member P_λ of a conformal one-parameter Lie group $\{P_\lambda\}_{\lambda \in \mathbb{R}}$ can be seen as a reparametrization of $\mathbb{R}^n$;

$$P_\lambda : \quad \begin{array}{ccc} \mathbb{R}^n & \longrightarrow & \mathbb{R}^n \\ x = (x_1, \cdots, x_n) & \longmapsto & (y_i(x, \lambda))_{i=1}^n \end{array}$$

characterized by

$$\left\| \frac{\partial P_\lambda}{\partial x_i} \right\| = \left\| \frac{\partial P_\lambda}{\partial x_j} \right\|$$

and

$$\langle \frac{\partial P_\lambda}{\partial x_i}, \frac{\partial P_\lambda}{\partial x_j} \rangle = 0 \quad \text{for } i \neq j, \ 1 \leq i, j \leq n .$$

In this event,

$$\sum_{i=1}^n \frac{\partial^2 \overline{v}}{\partial x_i^2} = 0 \implies \sum_{i=1}^n \frac{\partial^2 \overline{v}}{\partial y_i^2} = 0 .$$

When P_λ is linear, its action can be faithfully represented by an appropriate non-degenerate linear map $A : \mathbb{R}^n \to \mathbb{R}^n$, meaning $P_\lambda(x) = Ax \ \forall \, x \in \mathbb{R}^n$. Specifically, any transformation represented by a subgroup of the orthogonal group

$$O(n) := \{A \in \mathcal{M}_{n \times n}(\mathbb{R}) : A^T = A^{-1}\}$$

exhibits the required properties and these suffice for simplifying equation $(P2)$, provided that they leave the boundary condition invariant. Except for the case of $n = 2$, $O(n)$ has infinitely many one-parameter subgroups.

In addition to being conformal transformations, parametrizations by the orthogonal group are *isometries* meaning that

$$\left\|\frac{\partial P_\lambda}{\partial x_i}\right\| = 1$$

and

$$\|Ax\| = \|x\| \ ,$$

for every one-parameter subgroup $\{P_\lambda\}_{\lambda \in \mathbb{R}} \subseteq O(n)$ and $A \in O(n)$. Of course, we can deduce invariance of Laplace's equation (without boundary constraints) under the smaller rotational group $SO(n) \subseteq O(n)$.

For an illustration of this example, we will consider a simple solution of Laplace's equation, using the $n-$ball of radius a ; $\Omega = B^n(0,a)$ for convenience and the boundary constraint $h(x) = \varphi(\|x\|)$ on $\partial\Omega$. We take $\varphi : I \to \mathbb{R}$ to be a C^1 functional, while I is an open subinterval of the reals containing $\{a\}$. In this setting, any element of $O(n)$ leaves equation $(P2)$ invariant, and we can effectively deduce the Lie group invariant $r(x) = \|x\|$ which eliminates the group parameter λ since

$$r(P_\lambda(x)) = r(x) \ \ \forall \ \{P_\lambda\}_{\lambda \in \mathbb{R}} \subseteq O(n) \ .$$

We hereby seek a solution of functional form $\overline{v}(x) = \psi(r)$.

$$\frac{\partial \overline{v}}{\partial x_i} = \frac{d\psi}{dr} . \frac{\partial r}{\partial x_i}$$

$$= \frac{d\psi}{dr} . \frac{\partial\left[\left(\sum\limits_{i=1}^{n} x_i^2\right)^{\frac{1}{2}}\right]}{\partial x_i}$$

$$= \psi'(r) . \frac{x_i}{r}$$

$$\frac{\partial^2 \overline{v}}{\partial x_i^2} = \psi''(r) . \left(\frac{\partial r}{\partial x_i}\right)^2 + \psi'(r) . \frac{\partial^2 r}{\partial x_i^2}$$

$$= \psi''(r) \left(\frac{x_i}{r}\right)^2 + \psi'(r) . \frac{\partial}{\partial x_i}\left(\frac{x_i}{r}\right)$$

$$= \psi''(r) \left(\frac{x_i}{r}\right)^2 + \psi'(r) \left(\frac{r - \frac{x_i^2}{r}}{r^2}\right)$$

$$\sum_{i=1}^{n} \frac{\partial^2 \overline{v}}{\partial x_i^2} = \sum_{i=1}^{n} \left(\psi''(r)\left(\frac{x_i}{r}\right)^2 + \psi'(r)\left(\frac{1}{r} - \frac{x_i^2}{r^3}\right)\right)$$

$$= \psi''(r) + \psi'(r)\left(\frac{n-1}{r}\right)$$

$$\sum_{i=1}^{n} \frac{\partial^2 \overline{v}}{\partial x_i^2} = 0 \;\; \Rightarrow \;\; \psi''(r) = \psi'(r) \left(\frac{1-n}{r} \right)$$

$$\Rightarrow \int \frac{\psi''(r)}{\psi'(r)} dr = \int \left(\frac{1-n}{r} \right) dr$$

$$\Rightarrow ln(\psi'(r)) = (1-n)ln(r) + c$$

$$\Rightarrow \psi'(r) = \alpha r^{(1-n)}$$

where $\alpha \in \mathbb{R}$ is a constant of integration. For a further constant k of integration, we have solutions of

$\psi(r) = \alpha ln(r) + k$ for $n = 2$, and $\psi(r) = \alpha \dfrac{r^{(2-n)}}{2-n} + k$ for $n \geq 3$. Hence, the corresponding weak solutions in $H^1(B^n(0,a))$ are

$$\overline{v}(x) = \begin{cases} \alpha ln(||x||) + k & \text{for } n = 2 \\[2mm] \alpha \dfrac{||x||^{(2-n)}}{2-n} + k & \text{for } n \geq 3 \end{cases}$$

for $x \neq 0$. Due to uniqueness of the weak solution, it becomes clear that $(P2)$ often lacks a classical solution for this case, taking k to be zero and the function φ to be identity for instance [recalling $h(x) = \varphi(||x||)$ on $\partial\Omega$]. This is because $\overline{v}(x)$ as computed is not continuously extendable at $x = 0 \in B^n(0,a)$. Nevertheless, the above solutions $\overline{v}(x)$ are harmonic functions on $B^n(0,a)\backslash\{0\}$ and details about such functions are seen in classical potential theory.

4.4 Other Optimization Techniques

In many everyday applications, analogous quantified functionals are not differentiable, unlike the continuously differentiable examples considered above. In order to embrace a broader scope of functionals, we test instead for their lower semi-continuity and convexity. Given a functional

$$E : V \Rightarrow \mathbb{R} \cup \{+\infty\} \; ,$$

the domain D of E is the active region for our analysis;

$$D = \{v \in V : E(v) \in \mathbb{R}\}.$$

E is lower semi-continuous at $v_0 \in D$ if for every sequence $\{v_n\}_{n \in \mathbb{N}} \subset D$ which converges in norm to v_0, we have $\underset{n \to \infty}{liminf} \, E(v_n) \geq E(v_0)$. E is lower

semi-continuous on D if it is lower semi-continuous at every $v \in D$.
E is said to be convex if

$$E(\lambda u + (1 - \lambda)v) \leq \lambda E(u) + (1 - \lambda)E(v) \quad \forall u, v \in D \text{ and } \lambda \in [0, 1] \ .$$

If E is both convex and lower semi-continuous but not differentiable, then E is subdifferentiable and we invoke the Fenchel subdifferential of E at $u \in D$;

$$\partial E(u) = \{v^* \in V^* : \langle v^*, v - u \rangle \leq E(v) - E(u) \ \forall v \in V\} \ .$$

We have the subdifferential $\partial E(u) \neq \emptyset$ by subdifferentiability of E on D and elements of $\partial E(u)$ are called the subgradients of E at u. $\partial E(u)$ is always a convex set. If E is in addition differentiable at u, then $\partial E(u)$ is a singleton which coincides with the classical differential of E at u; $dE(u) \in V^*$. A necessary and sufficient condition for $\overline{v}$ to be a local minimizer of E is that $0 \in \partial E(\overline{v})$, and this is how we initiate weak formulations in this setting. As we have done in our illustrations, we may first consider the problem set in a larger reflexive space to make arguments about existence and uniqueness of solutions using the same given optimization theorem.

Epilogue

Riemannian geometry was the first branch of modern differential geometry to be developed, and it remains the most prevalent in practical terms for resolving diverse related problems. Though applications of differential geometry in pure mathematics lean heavily on the Riemannian branch, other branches such as the symplectic and Lorentzian have their irrevocable foothold in profound areas of tertiary mathematics and mathematical physics. Arguably, the most pronounced limitation in the incorporation of geometric techniques in computational problem-solving is the geometer's reliance on symmetries. Most atimes, in computations involving analytic manifolds without symmetry, comprehensive solutions cannot be attained. In fact, with the availability of axial symmetry alone, computations still tend to be quite cumbersome. This is the reason for choosing the sphere in $\mathbb{R}^n$ as the illustrative hypersurface for computation in this text; due to its radial symmetry but non-vanishing curvature.

Nevertheless, though the choice of the sphere eliminates some mathematical complications on one hand, it extends the array of prospective problems related to Riemannian geometry on the other hand. For instance, addressing Dirichlet's problem with the boundary constraint set on $aS^{n-1} \subset \mathbb{R}^n$ gave rise to the theory of Sobolev spaces in P.D.E's. This was because the fundamental solution as computed in chapter (4.3) was determined not to be continously extendable on the entire domain $B^n(0, a)$. Besides, potential theory and the theory of Dirichlet's functional are two overlapping but separately developed subjects stemming from Dirichlet's problem. For interested readers in touch with this text, details on how conformal Lie group theory is applied to assess the behavior of solutions to Dirichlet's problem based on the boundary constraint, should be an extract worth expounding. This involves point symmetries of the Laplace equation obtained via the prolongation technique for infinitesimal generators, as well as potential symmetries obtained via conservation laws of the equation.

Another instance whereby the sphere is involved in extrapolating theories from Riemannian geometry is the more modern Perelman entropy effect on 3-dimensional manifolds. Specifically, for a compact 3-manifold-without-boundary having positive scalar curvature, Perelman's entropy (with a modification to factor in manifold re-scalings) has the effect of metamorphosing the manifold into a 3-sphere with time. This is another application of Riemannain geometry worth expounding for interested readers. Perelman's entropy is an aspect of the Ricci flow, which is a potent contemporary mechanism for comprehending the topology and geometric analysis involved with intrinsic properties of Riemannian hypersurfaces and manifolds in general. Notably, the notion of symmetry groups is again relevant in the Ricci flow; as accommodated one-parameter groups of the underlying evolution equation $\left(\frac{\partial}{\partial t}g_{ij} = -2R_{ij}\right)$ characterize solitons of the flow.

Bibliography

[1] *DIFFERENTIAL GEOMETRY : CURVES - SURFACES - MANIFOLDS* - Wolfgang Kuhnel,
copyright 2003 by the American Mathematical Society.

[2] *APPLICABLE FUNCTIONAL ANALYSIS* - C. E. Chidume,
copyright 2006 by C. E. Chidume, ICTP, Trieste, Italy.

[3] *PARTIAL DIFFERENTIAL EQUATIONS* - Erich Miersemann,
October 2012 Department of Mathematics, Leipzig University.

[4] *GLOBAL ANALYSIS: DIFFERENTIAL FORMS IN ANALYSIS, GEOMETRY AND PHYSICS* - Ilka Agricola, Thomas Friedrich,
copyright 2002 by the American Mathematical Society.

[5] *CALCULUS OF VARIATIONS* - Jurgen Jost and Xiangqing Li-Jost,
copyright 1998 by the Cambridge University Press.

[6] *DIFFERENTIAL EQUATIONS AND GROUP METHODS FOR SCIENTISTS AND ENGINEERS* - James M. Hill,
copyright the CRC press.

[7] *TRANSFORMATION OF LINEAR DIFFERENTIAL EQUATIONS OF SECOND ORDER AND ADJOINED NONLINEAR EQUATIONS* - Lev M. Berkovich; Nikolai K. Rozov,
Archivum Mathematicum, copyright Masaryk University, Czech Republic, Vol. 33, No. 1-2, 75-98 (1997).

[8] *TRANSFORMATION AND FACTORIZATION OF SECOND ORDER LINEAR ORDINARY DIFFERENTIAL EQUATIONS AND ITS IMPLEMENTATION IN REDUCE* - Lev M. Berkovich; Fedor L. Berkovich,
University of Belgrade Publications (1995).

[9] *APPLICATIONS OF LIE GROUPS TO DIFFERENTIAL EQUATIONS (Second Edition)* - Peter J. Olver,
copyright 1993 by Springer - Verlag New York, Inc.

[10] *HAMILTON'S RICCI FLOW* - Bennet Chow, Peng Lu, Lei Ni,
copyright 2006 by the authors.

Printed by Books on Demand GmbH, Norderstedt / Germany